DISASTER PLANNING AND PREPAREDNESS: AN OUTLINE DISASTER CONTROL PLAN

British Library Cataloguing in Publication Data

Tregarthen Jenkin, Ian
Disaster planning and preparedness : an outline disaster control plan.—(British Library information guide, ISSN 0269-1809; v. 5)
1. Books——Great Britain——Conservation and restoration 2. Books——Great Britain ——Mutilation, defacement, etc.
I. Title
025.8'2'0941 Z701

ISBN 0-7123-3103-4

Ian Tregarthen Jenkin was born in 1920, went to Stowe and Trinity College, Cambridge, served in the Royal Artillery (1940-46), and studied painting at Camberwell and at the Slade, where he was subsequently Secretary from 1949 until 1975, when he returned to Camberwell as Principal. After retiring from there in 1985, he was appointed Curator at the Royal Academy Schools, but left in 1986 to become Director of the Open College of the Arts — for people wanting to study mainly at home and in their own time. A former member of the National Conservation Advisory Group, the Crafts Council, and the Art Panel of the Arts Council, he was appointed an OBE in 1984.

A companion to this guide is available as British Library R & D Report 5925, *Disaster planning and preparedness: a survey of practices and procedures.*

DISASTER PLANNING AND PREPAREDNESS: AN OUTLINE DISASTER CONTROL PLAN

IAN TREGARTHEN JENKIN

British Library Information Guide 5

Abstract

This guide is published in response to known demand among library and archive staff for some form of planning advice which could help in assessing the needs and problems of their own institutions and in preparing appropriate disaster control plans. It represents the second stage of a two-part research project, funded by the British Library and conducted at Camberwell School of Art & Crafts during 1984-85. The first stage was concerned with the nature and extent of disaster control planning in libraries and archives in England, Wales and Northern Ireland, and showed that 72% of the institutions which responded to the survey welcomed the prospect of guidance. The recommendations and procedures which follow include examples quoted from plans already in use; together with important additional information, they are arranged broadly in five parts:

(i) advice on ways of reducing the likelihood of disasters and on aspects of planning for their prevention;
(ii) matters for consideration, resolution and testing in advance, to ensure quick, orderly and correct responses when needed;
(iii) steps to be taken in the event of an actual disaster, including salvage and recovery procedures;
(iv) a bibliography;
(v) a range of general advice and information likely to be useful to library and archive staff and to others concerned in disaster control.

British Library Information Guides are published by the British Library Research and Development Department and distributed by the British Library Publications Sales Unit, Boston Spa, Wetherby, West Yorkshire LS23 7BQ. In the USA and Canada they are distributed by Longwood Publishing Group Inc, 27 South Main Street, Wolfeboro, New Hampshire 03894, USA, and in Japan by Kinokuniya Co Ltd, PO Box 55, Chitose, Tokyo 156.

ISBN 0 7123 3103 4
ISSN 0269 1809
SI/G/658

The opinions expresse... not necessarily those of the British Library.

Typeset by Type Out, London SW16 and printed in Great Britain at the University Press, Cambridge.

Contents

Acknowledgements

In expressing my gratitude to the British Library, and especially to its Research and Development Department whose encouragement and funding made this project possible, I would like to stress my particular indebtedness to the Director of its Preservation Service, David Clements, and convey to him my appreciation and thanks for all his help, wisdom, and patience as Chairman of the Project Steering Committee.

I wish to record, also, how very grateful I am to the individual members of the Steering Committee, for being so supportive during the past year, and for giving so generously in both time and advice.

My thanks go, too, to the large number of people, especially in public, university, and other libraries and archives, who, as individuals or through their institutions, provided so much assistance and supplied such valuable information. This is particularly so in the case of those who sent us their plans and agreed to extracts from them being used in this report.

Finally, I would like to thank my former colleagues in Camberwell and in the Inner London Education Authority, who, at a particularly difficult time in their history, so kindly undertook all the additional work which was imposed upon them in the course of this project.

Ian Tregarthen Jenkin
January 1986

1 Introduction

Disaster has been described by Fennelly[19] as '...an emergency event that occurs with little or no warning and that causes more destruction or disruption of operations than the museum can correct by application of its own resources'. This implies that, while internal staff should be able to take all necessary action in the case of any minor incident, the onset of disaster will call for urgent outside assistance.

The majority of incidents may be expected to result in water-related damage, and mould growth will normally start to develop on water-damaged material after approximately 48 hours. In planning to contend with disaster, therefore, the aim must always be to complete the salvaging of damaged holdings within this very short period of time.

Of equal importance with planning to contend with disaster is the need continually to reduce the likelihood of it happening. For this reason, practical preventative measures have to be worked out and perfected, to suit the needs and resources of each particular institution. If carefully planned and regularly monitored, these will substantially reduce the chances of an emergency and do much to contain the inevitable damage, should something disastrous actually occur.

This information guide is intended to meet known institutional demand for some form of flexible guidance, to aid the preparation of disaster control plans.

As speed and efficiency in action are of the greatest importance in dealing with disaster, it is essential to have in advance all necessary information and agreed procedures; ready access to emergency stores, equipment, services and advice; and, in relation to these matters, the absolute assurance of overall staff understanding and training.

The purpose of a disaster control plan is to provide for these things, and for their being known, rehearsed, and updated at all times and by all concerned.

Superior numbers in the text refer to the Bibliography, except in Appendix B where they refer to the numbered organisations in that section.

2 Background

A survey of the state of disaster planning in libraries and archives in England, Wales and Northern Ireland[50] showed that only 6.6% of the institutions surveyed had disaster control plans in use, and that a further 3.7% had plans in various stages of preparation. These included some of the institutions involved in extensive collaborative arrangements being made for extending to the Aberystwyth area the plans of the National Library of Wales.

The present publication is not intended to provide a further document on the lines already achieved in the National Library of Scotland *Planning manual for disaster control in Scottish libraries and record offices*[35]. Institutions concerned in the preparation of plans are strongly recommended to obtain copies of that manual. The information given here is intended only as a guide, in response to the demand revealed by the survey.

Many of the suggestions which follow have been written with the needs of smaller institutions very much in mind, and it is recognised that large organisations will probably have divisions of responsibility or established procedures which may well cause them to prefer different arrangements.

It has been assumed, however, that most institutions now considering the formulation of a disaster control plan will wish to nominate someone on the staff to take on the overall responsibility for disaster control. It is also assumed that, whatever the size of an institution, it will need to appoint from its staff a team (or teams) of appropriate people who are willing and able to take whatever action is necessary in response to a disaster. Should there be more than one such team, someone would have to co-ordinate their activities.

For the sake of simplicity, therefore, and because it may often have to be so in practice, it is further assumed that the Disaster Controller and the Disaster Team Co-ordinator (hereafter referred to as the DC) will in fact be the same person.

While it is believed that the various examples of planning which have been included will be of interest and help, it is emphasised that they are but extracts from much larger documents, designed to meet the overall needs and circumstances of specific institutions. As such, they show how part of a plan is concerned with some but not necessarily all of the points raised in a particular section of this report.

The extracts are taken from the following planning documents which are currently in use and which will be referred to by their abbreviations:

BLPS	British Library, Preservation Service: *Disaster control manual*, 1985
BMNH	British Museum (Natural History), Department of Library Services: *Planning for disaster control and salvage procedures*, 1985
NLS	National Library of Scotland: *Planning manual for disaster control in Scottish libraries and record offices*, 1985
OUWH	The Open University, Walton Hall, Offices and Library: *Safety code of practice*
SACG	Society of Archivists, Conservation Group: *Information needed for disaster planning*
UOLB	University of Oxford, Libraries Board: *The emergency treatment of water-damaged books*
USCW	University of Salford, Clifford Whitworth University Library: *Code of practice for emergencies*, 1983.

3 Prevention

3.1 General information

An appropriate member of staff should be given the responsibility of the DC.

Irrespective of the size of the institution, the first priority in planning for disaster control should be for an inspection to be made of the premises, inside and out, and of the immediately adjacent surroundings, with the idea of prevention very much in mind. This inspection should be made by the DC and by the member of staff responsible for security. There could be advantage in the same member of staff having responsibility for both these functions, as security is such an important factor in disaster control.

Some suggestions on the kinds of thing to be considered on that inspection are given in Appendix A, in the hope that they may be of help in smaller institutions not having staff who specialise in this type of activity.

The overall number of items which will need to be included under the heading of prevention in any plan will be determined by the circumstances of the institution concerned and by the nature and complexity of its holdings. There are, however, four cardinal matters which should be part of all plans: fire, flood, security, and storage.

3.2 Fire

It is expected that every institution will already have detailed fire regulations. These may, however, need to be extended, so as to include aspects of fire precautions, including particularly the procedures for ensuring staff awareness, and training in use, of the various types of fire-fighting equipment available in the institution. The survey showed that in only 37.5% of the institutions surveyed were staff trained in the use of such equipment[50]. It also showed that 10.6% of the institutions had not informed their local Fire Brigade about the building and its contents; that 38% had made no contact with the Chief Fire Officer; and that 15.6% held no periodic practice fire drills. Each of these factors should be included in institutional fire regulations. Valuable advice is available from the Fire Prevention Officer at the nearest Fire Brigade headquarters; and details of several fire safety organisations, and of the Fire Protection Association, are given in Appendices B and C. (See also Section 4.7 and Appendix A paragraphs 1.1, 1.3, 1.4, 2.2, 2.5, 2.6 and 4.2.)

3.3 Flood

Although most institutions will doubtless have a range of existing precautions to minimise the unwanted intrusion of water, additional measures may need to be considered, especially by those institutions which are in flood zones or consider themselves to be at risk from water or sewerage disruption. Analysis of the survey[50] showed that 16.8% of such institutions had material held in basements. Advice and help, particularly on the likelihood of disruption, is available from the local authority and from the appropriate water authority (see Appendix B). (See also Appendix A paragraphs 1, 2.3, 2.4, 3.3, 3.4 and 4.2.)

3.4 Security

As the whole matter of security is of great importance, and because such comprehensive advice on the subject is given in the NLS manual, the whole of the section there on security is reprinted below. Although some of its contents relate only to Scottish institutions, it is quoted in full because of its breadth of coverage. (See also Appendix A paragraph 2.7.)

NLS

SECURITY

Aim

To prevent damage arising from break-in when the institution is closed.

Comment

It is neither appropriate nor possible to include a fully comprehensive guide to building security in this manual. Each building poses its own problems of security and these are best discussed in detail with those responsible, e.g. University Works Departments, Local Authority Works Departments or the Crime Prevention Officer of the local police force. The purpose of this section, therefore, is to increase awareness of typical security problems and to make general recommendations.

Recommendations

A — Personnel
B — Equipment
C — General

A • Caretaker(s) and/or security guard(s) should check all entrances to building(s), including ground-floor windows and basements after closing-time each day and at least once in every 24 hours during holidays.

- Caretaker(s) and/or security guard(s) should enter building(s) out-of-hours at least once in every 24 hours to check vulnerable areas, e.g. computer rooms, photographic departments, conservation laboratories and workrooms, binderies, strong-rooms containing special collections etc.

- Where there is a caretaker or security guard he/she should be responsible for letting in cleaners, whether morning or evening, and should have a list of all cleaning staff. Any changes in cleaning staff should be notified to him/her immediately.

- Where there is no caretaker or security guard it is advisable for one trusted cleaner only to have a key to the building(s) and to be responsible for letting in any other cleaners.

- Any routine maintenance work taking place out-of-hours should be subject to surveillance by the caretaker(s) and/or security guard(s).

- Where there is no caretaker or security guard routine maintenance should not take place out-of-hours unless a senior member of staff is present.

- Emergency maintenance in- and out-of-hours should come within the sphere of disaster reaction.

B • Buildings with no caretaker(s) and/or security guard(s) should be fitted with an automatic intruder alarm system.

- All doors and windows should be fitted with strong locks.

- Fire exits which cannot be locked should be fitted with electronic alarms.

- Consideration should be given to fitting ground floor and sub-basement windows with toughened glass. Where windows have been broken by vandals or during an attempted break-in, toughened glass should always be used for replacement.

- All vandal damage should be repaired as quickly as possible. Swift reaction to such damage deters further vandalism as well as lessening the risk of a break-in.

- A careful check should be made of all keys issued to staff and an updated list kept of every key held by individual staff members.

- In cases where staff members do not take keys home, all keys left in a building should be locked away in a secure key-case sited in an unobtrusive position. The key to the key-case should either be handed to the caretaker or security guard, or taken home by a designated member of staff (with a deputy for sickness or holiday leave), or concealed in a place of safety well away from the key-case, preferably in another room.

- In cases where staff members do take keys home, they must be reminded to keep them safe at all times. Any loss of key(s) must be reported at once.

- Loss of a key or keys must be treated very seriously. If keys are not found within an agreed period of time consideration must be given to changing locks. In cases where an institution holds an insurance policy against break-in it is advisable to consult the insurers about continuation of cover in such circumstances.

C
- All staff members should be made aware of the need for good building security and should be asked to ensure that office windows, loading-bay doors etc, are firmly shut and, if necessary, locked before they leave each day.

- Staff members should be encouraged to be observant of open windows, unlocked doors, cracked or broken windows, damaged door catches and locks etc and report them.

- In buildings with no caretaker(s) or security guard(s), windows and doors in areas for which no individual staff member is responsible (e.g. lavatories and cloakrooms) should be checked by the buildings officer or other specially designated member of staff (or deputy).

3.5 Storage

The survey of disaster planning[50] showed that in almost half the number of buildings surveyed some part of the holding was kept below ground level; in 10% of the buildings it was more than half their range of material; and in 10.6% the entire range. While basement storage is likely to be avoided where possible, when it must be used much can be done to reduce the consequent risk to holdings. For detailed recommendations on storage, DCs should refer to BS 5454: 1977, *The storage and exhibition of archival documents*, and to BS 1153: 1975, *The processing and storage of silver-gelatin-type microfilm*. (See also Appendix A paragraph 3.)

3.6 An example of planning for prevention

The manner in which individual institutions will include in their plans those aspects of prevention which are important to them will vary according to their particular circumstances.

An extract from one document is given below, about this stage of planning, from a section headed 'Prior action'. The section concerned is not given in full, because it partly covers matters which, in this report, are considered later; further quotations from it are given then.

UOLB

THE EMERGENCY TREATMENT OF WATER-DAMAGED BOOKS

A Prior action

1 Introduction

Leaking water pipes, drains and roofs are the most common cause of disasters in local libraries. Fire is destructive but in many smaller library fires the pumping in of water causes very considerably more damage. Outbreaks of mould and infestations are also often a result of excessive humidity, absence of ventilation, and poor storage conditions. In most cases water remains the principal enemy. The following hints do not aim to be a complete manual for all occasions as circumstances, books and facilities vary greatly but they record what certain librarians have learnt from their own sad experiences. It is hoped that reading them may enable colleagues to avoid trouble, to prepare to minimise it, and to have some advice immediately to hand if it does occur. Section A consists of preventative and preparatory action, section B of emergency instructions.

2 Prevention

2.1 Premises

Flooding in libraries tends to occur either from the intrusion of rain water after heavy storms or, quite as often, from plumbing faults (e.g. central heating leaks; frozen cistern pipes or joins between pipes of different metals expanding at different rates; sprinkler systems). Librarians should consider all such problems on *or above* their own immediate premises. Bursars and administrators should be made aware of sources of danger and, after severe frosts or heavy storms (especially in the depth of the Long Vacation) all possible areas of threat (such as gutters, drainpipes and drains blocked by leaves, snow-covered flat roofs and parapets) should be checked as soon as possible. Basement areas, which are particularly liable to have water pipes passing near them or which may be affected by *rising* water (not forgetting outside flood water drain and sewer levels) are evidently particularly at risk and in them it is highly advisable to have the bottom shelf at least three to five centimetres off the ground. Sometimes a higher sill at the entrance to a basement room (say 150 mm.), although awkward to step over, may prevent passage water flowing in. Outlet drains can, where practicable, also be useful in basement areas. Bookcases against exterior walls should stand 1-2″ away both for ventilation and to keep any water flow clear of the shelves.

2.2 Fire extinguishers

At least in areas where rare or valuable books are kept it is highly advisable to provide dry powder fire extinguishers (or, equally suitable, asbestos fire blankets) rather than water extinguishers. Carbon dioxide extinguishers will cope effectively with small fires but one should note that the fire may re-ignite when the chemical has run out.

3 Preparation

3.1 Advance precautions

Consider stack cleanliness: water on top of dirt is more ruinous than water alone, as it may stain or abrade the pages and will also hold moisture, thus providing a breeding ground for mould spores. Consider the adequacy of shelf and catalogue records: the speed of emergency action and recovery can be seriously impeded by lack of up-to-date shelf marking, spine labelling and shelf records.

3.2 Unique items

In the case of small collections of unique or irreplaceable items not otherwise reproduced, boxing may provide some protection and microfilming may be a reasonable precaution against very severe damage or total loss. Microfilming can sometimes be arranged through local photographic units such as those of the Bodleian or the Department of Forestry. Always be aware that stamping dies may bleed and cause serious staining.

4 Preparation

4.1 General information

Having taken appropriate steps to reduce the chances of disaster, attention should be given to planning in advance all those arrangements which may have to be brought into action immediately in the event of an emergency.

Although the needs of institutions will differ, there are seven broad areas of preparation which ought to be considered:

i. procedures for out-of-hours emergencies
ii. formation and training of disaster team(s)
iii. in-house disaster control equipment or supplies
iv. back-up facilities, equipment and personnel
v. staff awareness, updating procedure, and early warning
vi. establishing salvage priorities
vii. regional assistance.

Examples of how these matters have been treated are again illustrated in extracts from the planning documents mentioned in Chapter 2.

4.2 Procedures for out-of-hours emergencies

As emergencies are likely to occur at night, or when institutions are closed or only partially manned, it is important to make arrangements for their detection at all times, and for summoning appropriate staff with the minimum of delay. Whatever the arrangement, it is necessary for anyone who may be first on the scene — e.g. cleaners, security staff, the police or possibly the Fire Brigade — to know whom they should contact on the institution's staff, and how.

Extracts are given from two documents:

1 UOLB

Emergency planning

Consider your emergency action and be sure that in your absence (e.g. on a holiday weekend) other people know what to do. Keep your emergency instructions (or these notes) easily available. Floods and their discovery are no choosers of time and indeed seem to prefer weekends and holiday periods. Keep telephone numbers of emergency key-holders or other staff (with home telephone numbers and arrangements to cover their absence, temporary or longer-term, in suitable places). Ensure that these are regularly updated, say every 1 July. Check emergency and out-of-hours access to water stopcocks and make sure they are clearly marked and their location known to a *number* of members of staff — this also applies to fuse boxes. All necessary interior keys must be available.

2 BLPS

PROCEDURES IN CLOSED HOURS

1 Reporting procedure

The following procedure will be used to report an emergency or incident occurring overnight, on Sundays or on public holidays when the Library is closed:

a. A security patrol or detection device will alert the Control Room of a fire, flood, or other source of damage.

b. The Control Room will send a party to the scene of the incident. They will try to contain it or put out any fire. Their report to the Control Room will determine whether the fire brigade is called or other action taken including contacting the Duty Officer.

c. The Duty Officer will contact the Disaster Assessment Officer from a nominated list.

d. In the event of a minor incident, the Duty Officer may initially discuss the situation by telephone with the Disaster Assessment Officer. They will decide whether the Disaster Assessment Officer need attend.

2 Response by Disaster Assessment Officer

a. The Disaster Assessment Officer will on arrival consult with the Duty Officer.

b. The Disaster Assessment Officer will now make a preliminary assessment and report back to nominated senior management. The Disaster Assessment Officer may now contact the Salvage Officer or may defer this until after more detailed assessment.

c. In the event of a minor incident the Disaster Assessment Officer may advise the Duty Officer by telephone at the reporting stage and will decide whether the damaged material requires immediate treatment or can be left until the next working day.

d. The Disaster Assessment Officer will decide in consultation with the Salvage Officer on the resources needed to handle the disaster.

e. The Salvage Officer will call out team members and inform them of the assembly point. If necessary the Disaster Assessment Officer may call out team members before the Salvage Officer arrives.

4.3 Formation and training of disaster teams

The needs of small institutions, and the range of options open to them, are clearly different from those of larger ones which have more staff available. Whatever its size, an institution needs to have its own in-house team or teams of people willing and able to be called upon, without warning, to undertake the immediate first-line measures in the event of disaster.

Again, two extracts are given:

1 NLS

ESTABLISHMENT OF TEAM(S) AND TRAINING

Aim

To form and train a team or teams of volunteer staff from all sections of the institution, who would be prepared to take part in salvage operations during or after a disaster.

Comment

When establishing a Disaster Reaction Team the size of the institution and the number of staff are critical factors. It is recognized that several institutions using this manual will have fewer than five staff members and specific recommendations for them are contained in C below.

In forming a team or teams it is important to decide on the member of staff (and deputy) who is to be in charge of disaster reaction. If there is only one team this person may be the team leader. If there is more than one team a Disaster Reaction Co-ordinator* must be appointed to take overall charge.

Recommendations

A — Recruiting
B — Training
C — Small institutions

A ● Recruit volunteer team members, bearing in mind that they should be
calm
adaptable
able-bodied
manually dextrous
able to get on well with colleagues
willing to work to strict guidelines.

● They should live within a reasonable distance of the institution by walking, driving or reliable public transport, and they should be available on the telephone.

● Inform heads of departments (and deputies) of the necessity for their advice and practical help in the event of a disaster affecting their particular holdings. List their work and home telephone numbers for immediate availability to the DRC and monitor holiday and sick leave to provide continuous cover.

● Form team(s) of 4-5 people ideally to include both curatorial and conservation skills. At least one team member should have a good working knowledge of the institution's holdings. The team leader (and deputy) should have some knowledge of conservation methods (*see* B).

B ● Train team(s) in disaster reaction procedures and familiarise each member with disaster prevention measures.

● Tour building(s) to point out

weak points in fabric
alarm systems — fire, intruder, water

* Hereafter the Disaster Reaction Team Leader or Co-ordinator is referred to as the DRC.

fire extinguishers
shut-off points for water, electricity, gas
location of disaster box(es)
location of telephone number lists of team members and heads of department (restricted access)
all entrances and exits
possible assembly points for teams in case of emergency.

- Arrange attendance of talk by local Fire Prevention Officer with showing of relevant film(s) e.g. fire extinguishing systems.

- Train team(s) in use of hand-held fire extinguishers and hosereels, as applicable.

- Train team(s) in handling of water- and fire-damaged books and records. Where no conservation expertise is available in-house or in a neighbouring institution seek specialist advice from the National Library of Scotland (NLS) or the Scottish Record Office (SRO).

- Issue each member of team(s) with a copy of your Disaster Control Plan.

- Keep all team members informed of any changes in your plan or developments in Disaster Control at regular meetings (at least once per annum).

C - If your institution has fewer than five staff members seek help from other institutions in the same or neighbouring buildings.

- Categories of people likely to be willing and helpful are

  Museum staff
  Library staff
  Local government office staff
  Schools staff
  University staff and students
  National Trust for Scotland volunteers.

- When making contingency arrangements choose volunteers carefully, bearing in mind that speed of reaction in a disaster may be hampered rather than aided by large numbers of willing but untrained helpers. This is particularly important in the case of a local record office or National Trust for Scotland property where the high percentage of irreplaceable material makes careful salvage essential.

2 *BLPS*

CONTACT LIST

1 Personnel

This list will contain the names, with phone numbers, of all disaster control personnel together with their addresses and home telephone numbers.

a) Disaster Assessment Officers: a nominated list with a rota showing first and second call.

b) Salvage Officers: named officers with a rota showing first and second call.

c) Disaster Control Team Members: personnel who can be contacted to attend a disaster following a fire, flood or other emergency. (Staff local to the area would be called on to assist in the event of a small scale incident.)

2 Duties of disaster control personnel

1 Disaster Assessment Officer
2 Salvage Officer
3 Disaster Control Team Members

Disaster Assessment Officer

One person from a nominated list to carry out the following duties:

1. To attend the scene of an emergency in working hours *only when authorised by the Control Room*;

   to attend the scene of a small scale incident in working hours when alerted on the Disaster Line;

   to liaise with the Duty Officer at all other times.

2. To take administrative responsibility for a disaster control operation, to call out the Salvage Officer as necessary, and to inform nominated senior management.

3. To liaise in an emergency with the fire brigade and other services.

4. To assess the scale of damage and to decide, in consultation with the Salvage Officer, on the resources (i.e. personnel, supplies and equipment, and facilities) needed to handle the disaster.

5. To keep the Disaster Control Zone Manual. This will contain the procedures, other information and plans outlined in this draft manual.

6. To write, after consultation with the Salvage Officer, a report for the Director of the Preservation Service on a disaster control operation.

Salvage Officer

One person from a list of named officers to carry out the following duties:

1. To accompany the Disaster Assessment Officer to the scene of a fire, flood, or other emergency in working hours *only when authorised by the Control Room*, and to attend at other times when called by the Disaster Assessment Officer.

2. To assist the Disaster Assessment Officer in deciding on the resources needed to handle the disaster.

3. To apply salvage procedures.

4. To supervise a disaster control team.

5. To contribute to the report by the Disaster Assessment Officer for the Director of the Preservation Service on disaster control operation.

Disaster Control Team Members

Staff will be called on to attend a disaster following a fire, flood or other emergency, and to carry out duties as instructed by the Salvage Officer.

Staff local to the area will be called on to assist in the event of a small scale incident.

4.4 In-house disaster control equipment and supplies

It is important to consider the stores, materials and equipment which will need to be kept locally available for immediate use in emergency. Time is essential, and such items can be hard to find at week-ends or on public holidays. The range and quantity of items needed will depend upon the size of the institution and the nature of its holdings. Such items should be kept in clearly marked disaster trolleys or boxes, and used only in time of emergency.

Extracts are given from four documents:

1 OULB

Emergency equipment

Consider your local availability of the more domestic equipment listed at the end of this pamphlet. Is it worth keeping to hand an emergency store of items such as clean brown paper, siliconized paper, freezer tape, etc? — these can be hard to find on Boxing Day. The Libraries Board has funded a general emergency store of such materials which is held by the Bodleian Library Conservation Section. In emergency and at closed times the section staff can be contacted through the Proctors' Office.

2 BMNH

Equipment and supplies

It is doubtful whether sufficient stocks of all salvage materials can be stored to deal with every emergency. However, substantial amounts of essential supplies will be located in all specialist libraries and clearly labelled. Larger items will be stored centrally with the Preservation Officer in the Archive Store or available for loan or hire at the time of emergency, through the Museum Accommodation Officer, DAS.

Supplies held by all libraries ('Disaster Box')

Plastic sheeting to cover tables
Black plastic rubbish bags for discarded material
Blotting paper for standing wet books on
Newsprint for interleaving and standing books on
Brown paper and tape
Paper towels
Jiffy bags
Nylon washing line
Crepe bandages
Small brush

Polymeter
Permanent fibre pens
Record forms

Equipment with Office Keeper and Housemen

Buckets
Mops
Sponges
Wet and dry vacuum cleaners
Overalls
Rubber gloves
Wellington boots

Available through the Museum Accommodation Officer, DAS

Plastic trunks or trays
Industrial fans
De-humidifiers
Freezer facilities
Thermohygrographs

3. SACG

In-house emergency equipment (attach a plan showing location of equipment)

A. Main utilities:

1. Main electrical cut-off switch
2. Main water shut-off valve/stopcock
3. Main gas shut-off
4. Sprinkler system/CO_2/halon system

B. Fire extinguishers (list types, numbers and users) and fire blankets

C. Smoke alarms

D. Portable pump (if on site)

E. Plastic sheeting (stored with scissors and tape)

F. Paper towel supply (if kept on site)

G. First aid kit

H. Cleaning utensils (buckets, mops, sponges etc.)

I. Industrial vacuum cleaner (wet and dry)

Have all members of Disaster Team toured A-I? ______________
Date

4 BLPS

EQUIPMENT AND SUPPLIES

1 Equipment will be kept in disaster control trolleys to be stationed on each book stack level, near Section Offices or known trouble spots. The trolleys will provide first aid when coping with a disaster while back-up supplies and equipment are held centrally.

2 *Standard equipment*

Each disaster control trolley will contain:

a. In storage space:

i. 1 roll of clingfilm, 300m × 45cm, for wrapping damaged materials
ii. 4 fold-up polypropylene crates, 36 × 6cm when folded
iii. 6 polythene sheets, 200 × 90cm
iv. 1 small plastic bucket with collapsible (i.e. telescopic) mop.

b. In drawers:

i. Stapler, staples, scissors, adhesive tape, permanent fibre pens and writing paper
ii. Melinex labels (translucent, white polyester) which can be written on when wet and stapled to wrapping
iii. Polythene bags in useful sizes
iv. Dusters, tissue paper, paper towels, torch
v. First aid box.

c. Plastic container for sawdust to fit on slide-out tray beneath drawers.

3. Back-up equipment

Central facilities will hold:

i. Electric water pump (in Binding Section sorting area of KEB)
ii. Water vacuum suction cleaner
iii. Fans, dehumidifiers
iv. Battery lights, torches
v. Blotting paper, with thymol solution to be made up when required
vi. Protective clothing: overalls, aprons, rubber boots and gloves.

4.5 Back-up facilities, equipment and personnel

All necessary back-up facilities should be located and contacted with care and precision. Preparatory plans should include the names, addresses and phone numbers of suppliers, their normal hours of availability, and details of whom to contact, and how, out of hours. It is important to establish, at the planning stage, the location, means of contact, and likely cost of the nearest available blast-freezing, freeze-drying, and microfilm-processing facilities; and of the Fire Brigade's pumping-out service, in case it may be needed even if there is no fire.

Extracts are given from one document:

SACG

OTHER SERVICES POSSIBLY NEEDED IN AN EMERGENCY

Service	Company and/or name of contact	Telephone number
Gas authority	....	....
Water authority	....	....
Electricity authority	....	....
Caretaker	....	....
Fire department	....	....
Fire alarm system	....	....
Police	....	....
Burglar alarm system	....	....

Service	Company and/or name of contact	Telephone number
Ambulance	....	....
First aider	....	....
Conservator in locality	....	....
Legal adviser	....	....
Electrician	....	....
Plumber	....	....
Carpenter	....	....
Glazier	....	....
Pest controller	....	....
Chemist	....	....
Mycologist	....	....
Locksmith	....	....
Architect/builder	....	....
Deep freeze facility	....	....
Others	....	....
	....	....
	....	....
	....	....

OFF-SITE EMERGENCY EQUIPMENT AND SUPPLIES

	Supplier (company or institution)	Telephone number
Hand trolleys	....	....
Drying racks	....	....
Refrigeration equipment	....	....
Dehumidifiers	....	....
Plastic containers	....	....

Portable electric sump pump	....	....
Heavy duty extension cables	....	....
Rolls of plastic sheeting	....	....
Electric (industrial) fans	....	....
Extra security personnel	....	....
Chemical suppliers	....	....
Unprinted newspaper	....	....
Dry ice	....	....
Freezer packaging materials (wax or silicone paper)	....	....
Plastic bin liners	....	....
Paper towels	....	....
Water hoses	....	....
Sponges, mops, buckets	....	....
Brooms	....	....
Gloves	....	....
Overalls	....	....
Boots	....	....
Facial masks/respirators	....	....
Tissue paper	....	....
Blotting paper	....	....
Folding tables	....	....
Aluminium foil	....	....
Adhesive tapes	....	....

Non-adhesive tapes	....	....
Torches	....	....
Vacuum cleaners	....	....
Waterproof pens/markers	....	....
Waterproof tie-on labels	....	....
Heaters	....	....
Emergency lighting	....	....
Others	....	....
	....	....
	....	....
	....	....

4.6 Staff awareness, updating procedure, and early warning

These three somewhat dissimilar matters are in fact closely related, because of the importance of early detection in the control of disaster, and the need for total staff awareness about agreed plans and procedures in emergency, about any updating that may have been introduced, and about any changes in the information about the provision of external services and supplies. It is therefore essential to devise a means of ensuring that all staff, especially new staff, are made fully aware of the potential hazards, what to look out for, and what to do if necessary. Arrangements should also be made for the regular and systematic updating of all information, monitoring of all established procedures, and notification of everyone concerned about any changes that are found to be necessary. (See also paragraph 5.2.)

Extracts are given from three documents:

1 OUWH

SAFETY TRAINING: INDUCTION CHECKLIST

This checklist is intended to cover those points which it is considered would be useful to a new employee on his/her first day at work.

It may be used in conjunction with the Open University Staff Handbook and the Information Services Department booklet 'An Introduction to The Open University'.

The two columns on the right can be used to record which information was given on the first day and subsequently.

SUBJECT	FIRST DAY	LATER
1. *Introduction*		
a. Welcome to the department/section	....	....
b. Brief description of the OU and its organisation	....	....
c. Description of departmental organisation	....	....
d. Introduction to staff	....	....
2. *Fire and safety*		
a. Tour of building	....	....
b. Fire alarm system	....	....
c. Fire exits	....	....
d. Fire evacuation procedure and assembly points	....	....
e. Location and use of fire-fighting equipment	....	....
f. Action in the event of fire	....	....
g. Fire notice/fire wardens/safety representatives	....	....
h. Safety rules	....	....
i. Accident prevention	....	....
j. Accident reporting	....	....

k. OU safety policy document . . . . . .

l. Requirements for protective clothing, footwear etc. . . . .

---

3. *General Welfare*

a. Location of medical centre . . . .

b. Facilities of medical centre . . . .

c. Registration of vehicles/parking discs/parking arrangements . . . .

d. Trade union representation and membership . . . .

---

4. *Introduction to section*

a. Introduction to manager, supervisor, colleagues . . . .

b. Introduction to work and layout of the section . . . .

c. Location of toilets, washrooms, cloakrooms, lockers, etc. . . . .

d. Brief explanation of main equipment and machines . . . .

e. Any particular fire or safety hazard . . . .

f. On-job training requirement . . . .

g. Place of work and section working arrangements . . . .

---

2 SACG

UPKEEP CHECKLIST — Date checked

A. Locks of doors and windows secure and keys accounted for? . . . .

B. Emergency numbers posted near every phone? . . . .

C. Last inspection by local fire department officials? . . . .

1. Fire extinguishers updated and operable? . . . .
2. Smoke alarms operable? . . . .
3. Sprinkler system operable? . . . .
4. Halon system operable? . . . .
5. Burglar alarm(s) operable? . . . .
6. Water detector operable? . . . .

D. Flashlights operable? . . . .

E. Staff familiarisation (by tour — not map) with location of fire extinguishers, flashlights and location of disaster teams . . . .

F. List of hazardous chemicals in-house (list to be sent to Fire Officer/Health & Safety Officer) . . . .

G. Last fire drill . . . .

H. Last inspection of external facilities . . . .

1. Roof checked after bad weather . . . .
2. Guttering checked after bad weather . . . .
3. Drains checked after bad weather . . . .
4. Loft/attic access points (external) . . . .
5. Loft/attic access points (internal) . . . .
6. Fire/emergency escapes (external) . . . .
7. Fire/emergency escapes (internal) . . . .
8. Windows checked for whether they will open . . . .
9. Fire hydrant . . . .

DISASTER TEAM RECORDS

Who on the staff has a copy of this plan and is familiar with its contents?

List all locations where this plan is on file (on and off the premises) and indicate provision of any further updates.

3 OULB

Early detection

Fire alarms have been known to draw attention to floods by short circuiting but in general most are found by human discovery. Inspection of infrequently used areas is therefore essential after storms or at other dangerous times, if not indeed on a regular basis. *Employees other than library staff must be encouraged to report danger signs* (damp patches, dripping taps, leaking radiators, musty smells) *near library areas* and the general administration should also inform the library of any humidity problems in its vicinity as soon as possible. Make sure that senior and other appropriate library staff are informed at once so that as much emergency action as possible is carried out by informed persons. Reporting should not preclude immediate action — if a radiator leak is soaking books *remove the books first, then report* the leak.

4.7 Establishing salvage priorities

At the time of a disaster, it will not be possible to start deciding what are to be the salvage priorities, or where the important, unique or irreplaceable holdings are located. Such decisions need to be taken on reflection and, therefore, well in advance. Having done this, arrangements should be made for the easy location and identification of such priority holdings, so as to maximise their chance of being salvaged. It may well be that, in case of extreme disaster, there is only the possibility of salvaging such material, and nothing more. The Chief Fire Officer at the local Fire Brigade should, therefore, also be informed of the nature and location of these items; the procedure adopted for doing so, and the information itself, being regularly monitored. If possible, disaster trolleys or boxes, fire extinguishers and other emergency apparatus should be positioned in the immediate vicinity of these holdings.

Extracts are given from two documents:

1 SACG

Priorities for salvage

Compile a list for each department and for each office of items which should be salvaged first following a disaster. The following considerations should be kept in mind when setting priorities:

A. Can the item be replaced? At what cost? (Cost figures should include ordering, cataloguing, etc., in addition to the purchase price.)
B. Would the cost of replacement be less or more than the cost of restoring/conserving the item?
C. How important is the item to the collection?
D. Is the item available elsewhere?

2 BMNH

Objectives of salvage and disaster procedures

To stabilize the condition of affected materials before removal for treatment by creating an environment which will as far as possible prevent further damage.

To salvage the maximum number of items from the damaged collections in a manner which will minimize future restoration and its costs.

Establishing priorities

Assessing the extent of the damage and establishing the priorities for salvage are two of the most essential components in a disaster strategy. Ask the question, 'How easy is it to replace this item and at what cost?'.

In broad terms, unique, rare and valuable items should take priority in a salvage situation.

Catalogues and records of holdings should also be rescued where possible.

For ease of identification of prime areas for salvage, all bookcases, cabinets and shelving containing materials of high priority are to be marked with a large white 'S' (Letraset or other transfer foil lettering).

In addition there should be a sketch plan of each library, held by the Salvage Officers, showing the location of priority material for salvage.

4.8 Regional assistance

It is of great importance to establish and maintain links with appropriate institutions, service organisations, conservation units, individual conservators, and other experts, particularly within the region. Help in identifying where such links might best be made could be sought from the various professional organisations or from the national libraries and record offices shown in Appendix B. Disaster control plans should include any useful information on such links, including the names of contacts, with telephone numbers, so as to ensure immediate access to someone concerned at any time. Reciprocal visits by some members of staff and, where appropriate, an exchange of plans should be arranged, in order to increase the prospects for mutual collaboration and support in the event of emergencies.

As all such planning will always be particular to the institution concerned, no example is given.

4.9 Categories of disaster or emergency

There is one further area of preparatory planning which may be thought appropriate in some of the large organisations with clearly defined administrative structures. In such institutions, it could be a useful aspect of planning to devise different procedures according to the category of emergency.

An extract is given from one document:

USCW

CODE OF PRACTICE FOR EMERGENCIES

1. *Introduction*

1.1 University policy on emergency situations

The policy of the University is that in any emergency the protection of staff, students and visitors is of primary importance. The protection of property is a secondary objective which must only be considered when all threats to life have been dealt with.

1.2 Types of emergency covered by the Code of Practice

Emergency situations covered by this code include:
- Fire and/or explosion
- Bomb threat
- Structural collapse of buildings or equipment
- Failure of services
- Flooding

Contamination by dangerous substances
Serious injury to persons
Riot or general disturbance

1.3 Application of the Code of Practice

Emergency situations may be considered as occurring at three levels of seriousness.

Major emergencies These are defined as those situations which *may* involve evacuation and/or closure of several buildings of the University. These incidents require central co-ordination of all actions being taken in individual buildings or areas.

Serious emergencies These are defined as follows:

(i) Those situations which *may* involve total or partial evacuations and/or closure of a single building, and/or

(ii) Those situations which require co-ordination of remedial action by various branches of Maintenance and Works and by individual departments.

All fire alarms will be classed as serious emergencies.

Minor emergencies These are defined as localised incidents which have no implications for the safety of persons in other parts of a building.

Minor emergencies are already dealt with on a departmental or sectional basis and it is not intended that the Code of Practice shall alter this procedure.

The intention of this Code of Practice and the associated guidance notes is to provide guidance for the effective handling of serious and major emergencies as defined above.

5 Reaction

5.1 General

Having made preparations to ensure that essential in-house and back-up arrangements are known by all concerned, and will be operable when necessary, the next stage in planning is to determine the various procedures that will have to be followed in dealing with an actual emergency.

There are six main components which need to be considered in all institutions:

— raising the alarm
— assembly of personnel
— stabilising the disaster area
— assessment of disaster
— salvage
— recovery.

5.2 Raising the alarm

In order that this may be done with minimum delay, it is essential that all staff, whatever their duties, should know precisely what to do, and when. New staff must, on appointment, therefore be made fully aware of the emergency arrangements, and the fact that these will be regularly tested and modified when necessary (see Section 4.6: Staff awareness, updating procedure, and early warning).

It is recommended that, in all institutions, alarm procedures should provide for the immediate summoning of the DC and of the Duty Officer or whoever is at the time in charge of the building.

One summary extract is given below, from plans for reporting on a disaster within working hours in a large institution. Although these may be too complex for use in many smaller organisations, they are useful as a general guide to content.

REPORTING PROCEDURE IN WORKING HOURS

1. *Fire*

Follow the fire drill in raising the alarm:

a. Raise the fire alarm by whatever appropriate means (e.g. break the glass on the fire call point).
Raising the fire alarm must take precedence over attacking a fire.

b. Providing there is no personal danger, a fire should be attacked with the first-aid fire-fighting equipment available.

2. *Emergencies other than fire*

a. Alert the Control Room of personal attack, theft, vandalism or damage threatened to the collections or other emergencies such as a suspected bomb by whatever system is provided in each building.

b. The Control Room will decide on appropriate action.

3. *Other incidents*

Other incidents, such as flooding from a burst pipe, may not warrant use of emergency procedures, and should be reported as follows:

a. During normal hours, the person discovering the incident reports the situation to the Admin Office on the Disaster Line.

b. At other times, the Duty Officer will be contacted.

c. The Admin Office will page the Disaster Assessment Officer and the Salvage Officer. Other personnel and facilities will be asked to stand by.

4. *Response by salvage personnel*

a. Fire

The Disaster Assessment Officer and the Salvage Officer will report in the first instance to the Assembly point.

They may not enter an area seriously affected by fire until the fire brigade has declared it safe.

b. Emergencies other than fire

The Disaster Assessment Officer and the Salvage Officer will report to the Assembly point and attend the scene when authorised.

c. Other incidents

The Disaster Assessment Officer and the Salvage Officer will go directly to the scene of the incident.

The Disaster Assessment Officer and the Salvage Officer will assess the scale of damage and the resources required, and will call out team members as necessary.

Staff working in the area can begin to move books out of danger pending the arrival of salvage personnel.

5.3 Assembly of personnel

Once the alarm has been given, appropriate personnel must be quickly assembled. The DC, having already been alerted as part of the alarm procedure, should:

i. decide on the need to summon the disaster team or teams and, if appropriate, any heads of department or other senior staff whose advice may be needed

ii. decide on a suitable assembly point for them

iii. summon those concerned and inform them of the chosen assembly point and, if necessary, how best to get there

iv. list all personnel who attend the emergency, and make this list available, if necessary, to the senior Fire Brigade officer present on the scene.

The BLPS extract given in 5.2 above provides for some of these procedures.

5.4 Stabilising the disaster area

In the event of disaster, the environment around it must be controlled and stabilised as quickly as possible, in order to minimise both the danger to life and the damage to the building and its holdings.

Mains service supplies (gas, water, electricity) must, as appropriate, be turned off. In the event of staff access being impossible, this will be done by emergency service personnel, who will probably require advice on the location of control points; this must be readily available.

In the event of fire, and to minimise water damage to the buildings, protective polythene sheeting from the disaster trolleys or boxes should be used if possible.

As soon as the area is declared safe to enter (by the emergency services, if present), and if possible while it is still being stabilised, the DC should, with members of the team(s) as necessary, proceed with the protection of any undamaged holdings. Damaged material should not be removed at this stage; an assessment of damage and a procedure for salvage must be determined first (see Section 5.5).

An extract is given from one document:

BLPS

STABILISING THE DISASTER AREA

1. After a major fire or flood the environment must be stabilised and controlled before and during the removal of damaged items. This is particularly important if salvage work is likely to last a considerable time.

2. The Disaster Assessment Officer must use discretion in applying the following procedures *whether or not damaged items are to be frozen*:

   a. Turn off all heat in the disaster area.

   b. Reduce relative humidity to 60% or less.
   Open doors. Open windows only if it is cold and not raining outside. Use fans, where necessary, to expel humid air. Dehumidifiers should be used only in small, enclosed areas.

c. If local electricity is out of order and cannot be quickly restored, use battery lamps and torches for lighting.

d. Put plastic sheets for protection over presses containing dry or undamaged materials.

e. Do not open wet books.
Do not disturb materials in archival storage boxes, protective boxes or acid-free envelopes.
Books immersed in water can be left until ready to move.

f. Do not remove damaged items until priorities have been agreed by the Disaster Assessment and Salvage Officers.

3. Small-scale incidents

The Disaster Assessment Officer should interpret the above procedures as is appropriate to the scale of the incident.

5.5 Assessment of disaster

As soon as any part of the disaster area has been declared safe (by the Fire Brigade's Officer in Charge, in the event of fire), and the protection of undamaged holdings has got under way, the DC should inspect as much as possible of the whole area (as permitted by emergency services, when present).

Where there is flood water, electricity supplies must first be turned off to avoid possibility of electric shock from 'live' water.

This inspection, if necessary in consultation with heads of department or other senior staff, should determine:

i. the nature and extent of the damage to the building and holdings

ii. whether flood water, if any, is contaminated and if so, from what source

iii. whether the holdings are secure from theft

iv. the resources likely to be wanted, and those still immediately available

v. the back-up facilities likely to be needed.

While making this inspection, the DC will also have to consider and start formulating an appropriate procedure for salvage (see Section 5.6).

An extract is given from one document:

BLPS

DISASTER ASSESSMENT

1. The Disaster Assessment Officer must first establish the kind(s) and scale of damage. Resources to handle the disaster must then be agreed with the Salvage Officer. In closed hours this consultation may take place over the telephone.

2. *Assessment of damage*

a. What kind of damage?

Burned materials will also be wetted by fire-fighting measures. Flooded materials will be damp, swollen and dirty. How severe is the damage?

b. How many items are affected?

Estimate their number approximately from the relative size of volumes and length of occupied shelving.

c. What kinds of material are affected?

Note the kinds of damaged bindings: vellum, leather, cloth, temporary, and estimate their relative numbers. Are they unbound printed materials? Loose materials in archival storage boxes? Photographic materials? Note rare or valuable items, such as the contents of a Pen, as priority items.

d. Is the building damaged?

Liaise with the Accommodation Officer to secure the disaster area for salvage work. The immediate area only may be sealed off. After a major fire or flood a large part or the whole of the area may be closed off after consultation with the Control Room.

e. Is the water in a flooded area contaminated?

Water may be contaminated by soot, ash, sewage or by having

passed through pipes or a gutter. If necessary, contact the Water Authority for advice. If tests show the water is contaminated, the Salvage Officer must decide whether to sterilise before or after salvage work.

f. Are the collections secure from theft?

Ensure that the collections are secure from theft when exposed during salvage work. There may be many outside personnel on site.

Boxes and crates must be accounted for before and after despatch to the binderies, conservation workshops and cold storage facilities. A member of staff may be required to accompany rare and valuable material going off-site to ensure its safety.

5.6 Salvage

This most crucial part of the plan is, in reality, likely to be successful if it has been accompanied by as much training of the team(s) and testing of the procedures as possible.

In the event of actual disaster, salvage should begin as soon as:

i. it is known to be safe to enter the area (with the permission of the Fire Brigade Officer in Charge, in the event of a fire)

ii. the DC has been able to complete a plan of action which will provide for:

   a. the selection of appropriate spaces for sorting, cleaning/treating, and packing damaged material

   b. the division of duties between the members of the team(s), and any others who may have been called in

   c. their relief, refreshment, and replacement as necessary.

Once salvage work has got under way, the DC should promote the arrangements for the transportation, blast-freezing and subsequent freeze-drying of severely damaged and packed material. (It is recognised that only 9.5% of the institutions which responded to the survey[50] had access to freeze-drying, and only 10.1% had access to cold storage, but it is strongly urged that all should seek such access.)

Damaged material should be quickly but systematically removed from the disaster area to the area chosen for sorting, where it should be divided into 'freeze' and 'non-freeze' categories. All photographic material, and items with vellum bindings, and all items that are damaged but not wet (or not very wet) should be put into 'non-freeze'. All others that are very wet should go into 'freeze'. 'Non-freeze' items should then be taken to the cleaning/treating area; 'freeze' items to the packing area.

Each item to be packed for freezing should be separately wrapped in polythene film or brown paper and marked with its shelving or other identification code; and an inventory should be kept to show briefly the code or pressmark, title, and details of content and materials for every item.

When the DC is satisfied that all necessary salvage has been completed and all damaged material despatched either for treatment or for freezing, cleaning up of the disaster area may be undertaken. Depending upon the state of the area, it may be necessary to remove to temporary storage all undamaged holding material until the building can be repaired and the environment rectified.

Extracts are given from two documents:

1 BLPS

SALVAGE PROCEDURES

1. Conditions in the disaster area must first be stabilised in order to make the building safe for salvage work. The collections must also be protected from possible theft or damage when exposed during salvage work.

2. Damaged materials must be removed from the disaster area. They must then be sorted and packed ready for transportation to the binderies, conservation workshops or freezer and, if necessary, for collection by outside cold storage firms.

3. Sort and pack in the nearest dry area.
   Handle wet or damp materials as follows:

   a. Handle quickly.

   b. Do not open wet books.

c. Do not close wet books open due to swelling.

d. Do not remove wet books from protective boxes or acid-free envelopes.

e. Do not disturb loose wet materials in archival storage boxes.

f. Do not clean or wash away dirt.

4 *Removal of damaged materials*

Remove damaged materials *in the condition in which they are found:*

a. Remove items first from the floor, then from between stacks. If there are no items on the floor, start where wettest.

b. Remove items next from presses, working down from the top. There will be a danger of collapse if lower shelves are cleared first.

c. Pass items by human chain to the packing area.

5 *Sorting damaged materials*

a. Sort items into *freeze* and *non-freeze.*

b. *Do not freeze:*
Vellum bindings and all documents consisting of vellum
Metal or tortoise-shell bindings
Seals and ostraka
Photographic and reprographic materials.

c. *Send for immediate freezing:*
Philatelic materials if they cannot be treated immediately: volumes will take priority over loose materials. Books consisting entirely of coated paper or containing plates on coated paper.

d. Sort for freezing *all other items that cannot be treated immediately.*

Sort in order of priority:

i. Materials which have already developed mould.

ii. Bound volumes: leather, cloth, temporary bindings.

iii. Archival storage boxes, unbound loose papers.

iv. Unbound materials, 'perfect' (i.e. adhesive) bindings.

6 *Packing damaged materials*

Wrap only items to be collected for freezing by outside cold storage firms.

Do not wrap items to be sent direct to the binderies, the conservation workshops or the freezer.

a. Wrap each item in a polythene bag or film cut from a roll.

b. Write pressmark on Melinex label and staple at top.

c. Pack small to medium-size items fairly tight in fold-up crates from nearby disaster control trolleys.
Lay books flat to avoid distortion.
Do not stand books on their fore-edge. Books containing coated paper should stand vertically to avoid compression.

d. Lay large items flat in non-collapsible crates.

e. Load crates to 10-15 kilos (20-30lb) and number each crate.

f. Philatelic collections
Lay volumes flat. Take care not to dislodge stamps from their mounts.
Loosely wrap each item in a polythene bag or film cut from a roll.
Liaise with Philatelic staff as necessary.

g. Reprographic materials
Wet microfilm will stick together if allowed to dry. Place in polythene bags and store in water-filled containers.
Wet microfiche will also stick. Lay flat to dry on white blotting paper.
Liaise with Reprographic as necessary. They will arrange laboratory treatment if required.

h. Photographic materials
Unexposed film will stick together if wet. It can be dried only if sealed.
Do not lay wet film flat. Put on hangers in a drying cabinet.

Liaise with Photographers as necessary. They will advise on suitable treatment.

7 Transportation

Crates of material must be checked before and after despatch. A member of staff may be required to accompany rare and valuable material going off-site to ensure its safety.

a. Despatch items for immediate treatment in crates to the binderies and conservation workshops.

b. Despatch items to be frozen in crates to the freezer.

c. Arrange collection of crates containing items for freezing by outside cold storage firms from the loading bay or forecourt.

8 Cleaning up

When salvage work is completed, the disaster area must be cleaned and aired. Structural repairs may be necessary. Liaise with the Accommodation Officer as necessary.

Materials not directly affected may contain excess moisture. Outhousing may be necessary where materials can be kept under controlled conditions until the disaster area has been renovated. Liaise with Public Services, Storage and Delivery as necessary.

2 BMNH

Freezing books

1. If damage is very extensive or books are too fragile to dry by air-ventilation, they can be deep-frozen. Freezing books that are water-soaked to between approximately −18 °C and −20 °C halts deterioration until drying facilities are available. Books may remain frozen indefinitely without harm.

2. If possible, books for freezing should be wrapped singly in brown paper or placed in Jiffy bags and marked with a record number. Each book should be listed on a record form [Figure 1] as it may be some time before they are treated. They should then be packed loosely in plastic trunks or boxes for transportation to refrigeration facilities.

3. If the quantity of water-soaked books is so great that wrapping and numbering is not feasible, books may be placed in polythene bags, or directly into plastic trays, trunks or cardboard boxes for freezing.

4. It is important to try to maintain some kind of record, however brief, of material to be frozen as facilities will almost certainly be distant from the BM(NH).

5. If facilities are available, frozen material may be vacuum freeze-dried at a later date. This process is more likely to result in bindings being saved. Otherwise conventional drying methods are used when time permits.

BRITISH MUSEUM (NATURAL HISTORY)
DEPARTMENT OF LIBRARY SERVICES

INVENTORY OF WATER-SOAKED BOOKS FOR FREEZING

Library: Date wrapped for freezing: Location of freezer facilities:

Was water: clean? ☐ dirty? ☐ contaminated? ☐

Item No.	Pressmark	Author/Title	manuscript	pencil or w.c. drawing	vellum	leather	cloth	paper covers	coated paper	colour plates	maps	photographic	Notes

(tick boxes)

Figure 1: Inventory of water-soaked books for freezing

5.7 Recovery

Items removed for freezing will be treated later, their deterioration meanwhile being halted by the freezing process. All other items will need to be dried by air-ventilation immediately.

The following extract (which is based on *Procedures for salvage of water-damaged library materials,* 1979 by P Waters[52]) gives guidelines for drying by air-ventilation, for treating mould, and for dealing with specific types of material.

BMNH

Drying books by air-ventilation

1. If books are strong enough, stand them on their heads (i.e. upside down) with their covers only opened slightly, on an absorbent material such as blotting paper or unprinted newsprint.

2. Change the absorbent paper frequently to keep humidity levels low.

3. Books which cannot stand can be leant lightly against one another or placed spine upwards at a shallow angle of opening, resting on the unopened pages.

4. When books are dry, they should be closed up, laid flat on a table, and gently formed into the normal shape, with convex spine and concave fore-edge, and held in place with a light weight for several days.

5. Books are likely to require some days to dry, and the slower the process the more likely it is that volumes will return to something like their original condition.

Mould

1. Temperature and humidity in areas where books are drying should be kept within the range 15-20 °C and 50-60%. Good ventilation will reduce the risk of mould growth.

2. If mould appears outside or inside books, urgent action is essential.

3. Isolate affected volumes.

4. Loose mould can be gently removed using a small, soft paint brush, either outside or in a well-ventilated room.

5. Bindings can be wiped using a 2% thymol/alcohol solution on a tissue. The thymol solution may be obtained from the Mineral Laboratory.

6. Pages should be interleaved with thymol-impregnated sheets. These sheets consist of an absorbent paper such as blank newsprint which has been coated with a 10-15% solution of thymol crystals in ethanol or acetone, and allowed to dry.

7. The thymol sheets will be prepared either by the Mineral Laboratory or by a private conservator. Salvage Officers should liaise with them.

8. Do not over-interleave a volume. Distortion can occur and bindings could break. Every ⅛ of an inch throughout the text is sufficient.

Guidelines for specific types of material

1. Books

1. Any type, damp or partially wet — dry by air-ventilation.
2. Modern, soaked — dry by air-ventilation; freeze if rare or valuable.
3. Pre-1800, soaked — freeze unless expert help available immediately.
4. Leather-bound — freeze as 3 above.
5. Vellum-bound — freeze as 3 above; bind with crêpe bandages before freezing to keep volume in correct shape.
6. Coated paper (i.e. with high gloss finish) — keep wet until it can be interleaved with polythene or melinex; freeze if time is limited.
7. Covered in mud or dirt — rinse under clean running water, holding the book firmly closed.
8. Hand-coloured illustrations — freeze unless expert help available immediately.

2. Manuscripts, drawings and loose sheets

1. Boxes, portfolios, envelopes etc. often protect manuscripts and drawings from most of the damage. If it is only the container that is wet, remove contents and if necessary discard boxes. Wrap the contents in clean white paper or envelopes, which should be clearly marked in pencil.

2. Partially wet or damp: dry between blotting paper, using weights to flatten sheets when almost dry. Pamphlets may be hung over plastic washing lines.

3. Water-soaked: wrap in blank newsprint or white paper and freeze. Do not attempt to remove them from their box if they are soaked right through.

3. Maps

1. Map cabinets may well protect maps from the worst damage, but all maps should be checked for signs of wetness or mould.

2. Partially wet or damp: spread the maps out flat on tables on clean blotting paper; use weights to keep them flat when almost dry.

3. If the linen backing starts to peel off because of water-soluble glue, lay maps face down on blotting paper, or face up on polythene to prevent the glue sticking.

4. Water-soaked: dry by air-ventilation, laid flat on absorbent paper. If too many to lay out singly, stack in small piles changing blotting paper interleaving regularly.

4. Photographic material

1. Diazo copies of microfiche are not worth rescuing if they are totally soaked.

2. Negatives, prints, transparencies and silver halide microfilm: damp or partially wet — dry by air-ventilation by hanging up or laying flat on blotting paper.

3. Negatives, prints, transparencies and silver halide microfilm: water-soaked — place in clean *cold* water in polythene bags or buckets and wait for expert help. If expert help is not available within 48 hours, soaked photographic material must be frozen. It can then be thawed and dried when time permits.

4. Glass plates and lantern slides: dry by air-ventilation. They should not be immersed in water or frozen.

5. Photograph albums — dry by air-ventilation if only damp; wrap and freeze if soaked.

Subsequent action

1. Storage areas must be carefully cleaned and dried before any books are returned.

2. Survey all library storage constantly for signs of mould.

3. Prepare a written report of the disaster and salvage procedures, pointing out areas where improvements can be made and lessons learnt, and assessing total damage.

Conclusion

Disasters tend to be very individual things; two are rarely the same. There are certain general rules but circumstances will vary enormously. The use of common sense is probably the most important rule of all.

Bibliography

1. Allen, Stephen J. Crisis public relations communications. In: *Museum, archive and library security*, ed. L J Fennelly. Boston and London, Butterworth, 1983, pp 131-61.

2. Baker, J P and Soroka, M C, eds. *Library conservation: preservation in perspective.* Stroudsburg, Pa., Dowden, Hutchinson and Ross, 1978.

   Section IX, pp 383-408, covers disasters and salvage.

3. Baker, R and others. *A disaster plan for the American Antiquarian Society.* Worcester, Mass., AAS, 1983.

4. Banks, P N. *A selective bibliography on the conservation of research library materials.* Chicago, Society of American Archivists, 1981.

5. Bannister, J E and Bawcutt, P A. *Practical risk management.* London, Witherby, 1981.

6. Barber, G. Noah's ark, or, thoughts before and after the flood. *Archives*, 16(70), October 1983, 151-61.

7. Bohem, H. *Disaster prevention and disaster preparedness.* Berkeley, Cal., University of California Task Group on the Preservation of Library Materials, 1978.

8. British Standards Institution. *Recommendations for the processing and storage of silver-gelatin-type microfilm.* (BS 1153: 1975.) London, BSI, 1975.

9. British Standards Institution. *Recommendations for the storage and exhibition of archival documents.* (BS 5454: 1977 (currently under revision).) London, BSI, 1977.

10. Buchanan, S and others. *The Stanford Lockheed Meyer Library flood.* Stanford, Cal., Stanford University Libraries, 1980.

11. Chemical Industry Safety and Health Council: Safety Committee. *Recommended procedures for handling major emergencies.*

12. Cockerell, H A L. *Witherby's dictionary of insurance.* London, Witherby, 1980.

13. Council of Federal Libraries. *Library disaster preparedness.* (Information series 8.) Ottawa, National Library of Canada, 1984.

14. Crockford, N. *An introduction to risk management.* Cambridge, Woodhead-Faulkner, 1980.

15. Cunha, G M. An evaluation of recent developments for the mass drying of books. In: *Preservation of paper and textiles of historic and artistic value*, ed. J C Williams. (Advances in Chemistry 164.) Washington, DC, American Chemical Society, 1977, pp 95-104.

16. Cunha, G M. *What an institution can do to survey its conservation needs.* New York, NY, New York Library Association, 1979.

17. Cunha, G M and Cunha, D G. *Library and archives conservation: 1980s and beyond.* 2 vols. Metuchen, NJ, and London, Scarecrow Press, 1983.

    Volume 1, Chapter 7 deals with disasters.

18. Feilden, B. Museum management and natural disasters. *International Journal of Museum Management and Curatorship*, 1, 1982, pp 231-35.

19. Fennelly, L J, ed. *Museum, archive and library security.* Boston and London, Butterworth, 1983.

20. Fischer, D J. Problems encountered, Hurricane Agnes flood, June 23rd 1972, at Corning, New York, and the Corning Museum of Glass. In: *Conservation administration*. North Andover, Mass., New England Document Conservation Center, 1975, pp 170-87.

21. Fortson-Jones, J. *Manual for paper preservation and disaster planning for archives, libraries and museums in Nebraska.* Nebraska City, Nebraska State Historical Society, 1981.

22. Gibson, J A and Reay, D. *Drying rare old books soaked by flood water.* Harwell, Atomic Energy Research Establishment, 1980.
    (A summary appears in *Museums Journal*, 80(3), December 1980, pp 147-48.)

23. Harvard University Library: Subcommittee on Emergency Procedures. *Emergency guidelines for Harvard libraries.* Cambridge, Mass., Harvard University, 1982.

24. HM Factory Inspectorate: Accident Prevention Advisory Unit. *Effective health and safety policies: a review of the work of the Accident Prevention Advisory Unit.* London, HMSO.

25. Home Office. *Safe use and storage of liquefied petroleum gas in residential premises.* (Fire Prevention Guide 4.) London, HMSO, 1976.

26. Hunter, J E. *Emergency preparedness for museums, historic sites and archives: an annotated bibliography.* Nashville, Tenn., American Association for State and Local History, 1979.

27. International Federation of Library Associations. Principles of conservation and restauration in libraries. *IFLA Journal*, 5, 1979, pp 292-300.

28. Lowell, H P. *Planning for library conservation: a needs assessment manual.* Denver, Col., Colorado State Library, 1981.

29. Martin, J H. Après le déluge — resuscitating a water-logged library. *Wilson Library Bulletin*, 50, November 1975, pp 233-41.

30. Moor, I and Moor, A. *Conserving photographs.* London, Bishopsgate Press, 1985.

31. Morris, J. *Managing the library fire risk.* Berkeley, Cal., University of California Office of Insurance and Risk Management, 1981.

32. Morrow, C Clark. Disaster preparedness. In: *The preservation challenge*, ed. C C Morrow and G Walter. White Plains, NY, Knowledge Industry Publications, 1983, pp 73-77.

33. Myers, J N and Bedford, D D, eds. *Disasters: prevention and coping: proceedings of the conference, 21-22 May 1980.* Stanford, Cal., Stanford University Libraries, 1981.

34. National Fire Protection Association. *Protection of libraries and library collections.* Boston, Mass., NFPA, 1980.

35. National Library of Scotland. *Planning manual for disaster control in Scottish libraries and record offices.* Edinburgh, National Library of Scotland, 1985.

36. New York University Libraries: Preservation Committee. *Disaster plan workbook.* New York, NY, New York University Libraries, 1984.

37. Northwestern University Library. *The memorial day steam cleaning.* Evanston, Ill., Northwestern University, 1977.

38. Oklahoma Chapter of the Western Conservation Congress 1981-82: Disaster Task Force Committee. *Basic guidelines for disaster planning.* Oklahoma City, Department of Libraries, Archives and Records Division, 1982.

39. Price, Robin. Preparing for disaster. *Journal of the Society of Archivists,* 7(3), April 1983, pp 167-72.

    Text of paper given at the International Meeting for Archive Conservators, September 1982.

40. Ruggere, C and Morse, E H. The recovery of water-damaged books at the College of Physicians of Philadelphia. *Library and Archival Security,* 3(3/4), Fall/Winter 1980, pp 23-28.

41. Rutstein, J S. *Preservation and collection development: establishing the connection.* Paper presented at the ACRL National Conference, Minneapolis, October 1981.

42. Schmelzer, M. Fire and water: book salvage in New York and Florence. *Special Libraries,* 59(8), October 1968, pp 620-25.

43. Society of American Archivists. *Basic archival conservation program conservation self study.* Chicago, SAA, 1982.

44. Society of American Archivists. *Disaster prevention and preparedness.* (Problems in Archives Kits.) Chicago, SAA, 1982.

45. Spawn, W. After the water comes. In: *Library conservation,* ed. J P Baker and M C Soroka. Stroudsburg, Pa., Dowden, Hutchinson and Ross, 1978, pp 400-08.

46. Swartzburg, S G. Disaster planning. In: *Conservation in the library*, ed. S G Swartzburg. London, Aldwych Press, 1983, pp 18-25.

47. Systems and Procedures Exchange Center (SPEC). *Preparing for emergencies and disasters.* Washington, DC, Association of Research Libraries.

48. Thomas, D and Flink, J M. Rapid drying of water-soaked books using a microwave tunnel dryer. *Restaurator*, 2(2), 1975, pp 105-19.

49. Tillotson, R G. *Museum security: la sécurité dans les musées.* Paris, International Council of Museums, 1977.

50. Tregarthen Jenkin, I E. *Disaster planning and preparedness: a survey of practices and procedures.* (British Library Research and Development Report 5925.) London, British Library, 1986.

51. Upton, M S and Pearson, C. *Disaster planning and emergency treatment in museums, art galleries, libraries, archives, and allied institutions.* Canberra, Institute for the Conservation of Cultural Material, 1978.

52. Waters, Peter. *Procedures for salvage of water-damaged library materials.* 2nd ed. Washington, DC, Library of Congress, 1979.

Appendix A — Disaster Controller's inspection of the building

It is suggested that, taking a plan of the building, or making a provisional one, *en route,* if necessary, the DC should go methodically round the building, inside and out, to observe and make notes on anything which could permit, encourage, or inhibit unwanted occurrences.

1 *External*

1.1 The outside of the building and its surrounding area and premises should be most carefully considered, with particular attention being given to any neighbouring activities which might, for example, be sources of potential fire hazard. Contact with the local Fire Brigade Chief Officer is important at this stage.

1.2 Potential flooding problems should be looked for, including the possibility of water or sewerage disruption. Contact with the local authority and area Water Board is likely to be valuable in this respect, especially regarding the locality of flood zones.

1.3 The exact location of manhole covers and of gas, electricity, and water input and cut-off points should be indicated on the plan, and discussed if necessary with the services concerned.

1.4 The building's overall accessibility for vehicles or equipment in times of emergency must be considered, and discussed with the local Fire Brigade, if thought necessary.

1.5 Attention should be given to the state of doors, windows, skylights and roofing. If access to the roof is possible from within, the roof should also be inspected for any signs of intrusion by water. (In this connection, the DC will want to ensure that roof areas are inspected during and after heavy storms.)

1.6 Outside walls, below any damp-proof coursing, should be checked to ensure that undesirable material has not collected above that level. Underfloor ventilators should at the same time be inspected to see that they are functioning properly.

1.7 The height of thresholds at ground-floor and basement levels should be checked and raised where necessary to deter flooding within the building.

1.8 Finally, all drains, downpipes, guttering and gulleys should be inspected, to ensure that there is no damage or blockage causing any potentially hazardous condition.

2 *Internal*

2.1 The general layout of accommodation and the means of access or escape should be noted on every floor.

2.2 The type and location of alarm systems and of fire-fighting and other emergency equipment should be noted and their positions recorded on the plan. Their most recent date of testing should also be noted at the same time, so that regular testing may later be introduced if it is not already done.

2.3 Careful inspection should be made of any catering, cloakroom or lavatory provision, and of drains, gulleys, water pipes and stopcocks, especially those which may be above or adjacent to storage/holding areas. The location of stopcocks to control all water runways should be noted and indicated on the plan.

2.4 Out-of-hours or emergency arrangements for providing access to stopcocks, at all times, should be considered and if necessary perfected. As the importance of stopcocks, in contributing to the control of disaster, cannot be over-emphasised, the need and possibility of increasing their provision should be given high priority. In any case, the exact location and function of each one should be shown on the plan.

2.5 The state and age of all electrical wiring and apparatus should be considered, and a record made of when re-wiring was last carried out. (As far as buildings are concerned, electricity boards normally recommend re-wiring every 25 years.)

2.6 Arrangements for the prohibition of smoking throughout the building should be considered. Special provision may need to be made for contractors' employees, who may therefore be seen as presenting particular risks. If smoking has to be allowed in specified areas, the supply of sand trays at their entrances and exits should also be considered.

2.7 Security arrangements should be noted, for the control of staff, contractors' employees, members of the public, and authorised users, both on entering and on leaving the building. While it would be impossible to exclude a really determined intruder, or to prevent a carefully conducted theft, much can be done to reduce the chances of these things happening and their likelihood of success.

3 *The collection*

3.1 The DC needs to be informed of the general contents of each room, and of any irreplaceable items of unique importance, and should arrange for updated information on these things to be regularly made available.

3.2 In addition to giving special attention to the storage and security of irreplaceable items, consideration should be given to ensuring that their text and form are recorded on microfilm or otherwise, and that such duplicate records are kept in separate custody.

3.3 Care should be taken to ensure that unlined bookcases are never positioned against outer walls, particularly in basements.

3.4 Throughout the collection, any stamping of ownership or marks of identification should be done only with the use of waterproof material; soluble ink or dye will cause serious harm with humidity.

3.5 Particular attention should be given to the general state of cleanliness of bookstacks and other storage areas. Humidity and dirt together present a potentially perfect breeding ground for mould spores in addition to causing staining.

4 *General*

4.1 After the survey, when any potential hazards will have been identified, an in-house preventive programme should be developed by the DC, and any necessary precautions taken or improvements carried out, to the satisfaction of the DC.

4.2 It is strongly urged that, if the following are not already provided for, the DC should take steps to arrange them:

- i. the regular training of staff in the use of fire-fighting apparatus and in First Aid
- ii. the regular testing of fire extinguishers
- iii. the introduction of fire-drills
- iv. regular contact between the DC and the Chief Fire Officer
- v. regular updating of information given to the Fire Brigade on the building and contents
- vi. re-wiring of the building, if necessary, preferably about every 25 years
- vii. access to equipment for treating damaged books or documents
- viii. access to freeze-drying, cold storage, and dehumidifiers
- vix. the maximum possible amount of air-conditioning and humidity control.

Appendix B — Useful organisations and institutions

Contents *Page*

Superior numbers in the text refer to other organisations in this section.

1 AREA MUSEUMS SERVICE

A membership organisation, consisting of representatives of museums and the other organisations which run them, with the object of helping local museums to improve the standard of care for their collections and service to the public. This is done by fostering and increasing co-operation, providing common services and information, and distributing government funds to approved projects.

Area Museum and Art Gallery Service for Yorkshire and Humberside	Farnley Hall, Hall Lane, Leeds LS12 5HA (0532) 638909
Area Museum Council for the South West	Hestercombe House, Cheddon Fitzpaine, Taunton, Somerset TA2 8LQ (0823) 59696
Area Museums Service for South East England	34 Burners Lane, Kiln Farm, Milton Keynes, Bucks MK11 3HB (0908) 563997
Council for the Museums in Wales	32 Park Place, Cardiff CF1 3BA (0222) 25432
East Midlands Area Museum Service	Courtyard Buildings, Wollaton Park, Nottingham NG8 2AE (0602) 288749
North of England Museums Ltd	27 Sutton Street, Durham D1 4BW (0385) 46665
North Western Museum and Art Gallery Service	Griffin Lodge, Griffin Park, Cavendish Place, Blackburn, Lancs BB2 2PN (0254) 670211
Scottish Museums Council	County House, 20-22 Torphichen Street, Edinburgh EH3 8JB (031) 299 7465

West Midlands Area Museum Service	Avoncroft Museum of Buildings, Stoke Heath, Bromsgrove, Worcs B60 4JL (0527) 72258

2 ASBESTOS INFORMATION CENTRE, LTD
St Andrew's House, 22-28 High Street, Epsom, Surrey KT19 8AH
(03727) 42055

Founded in 1967, by leading members of the British asbestos industry, to provide information and an advisory service on asbestos. Holds meetings with government and unions, gives technical advice on asbestos and safety, provides a range of literature, and is willing to answer questions about the identification, use or removal of asbestos in libraries and archives.

3 ASSOCIATION OF BRITISH INSURERS (ABI)
Aldermary House, Queen Street, London EC4N 1TU
(01) 248 4477

The recent formation of this new insurance body is, as a development, related to the creation of a new Loss Prevention Council[22] of which, with Lloyd's, ABI has become a corporate member. It is represented on the Home Office Standing Committee on Crime Prevention and maintains links with the police and other organisations concerned in security through the Home Office Crime Prevention Panels. It is closely involved with the work of the National Supervisory Council for Intruder Alarms[26] and has representatives on several committees of the British Standards Institution[8] concerned with security. As a consequence of the developments associated with its formation, the administration of many of the interests of the Fire Offices' Committee[14] has been incorporated into the ABI. Producing a wide range of crime prevention publicity, regular theft-loss statistics, and information about theft insurance and prevention, the Association is willing to give general advice on such insurance matters to library and archive staff.

4 BRITISH AUTOMATIC SPRINKLER ASSOCIATION, LTD
(BASA) 10 Chellaston Road, Wigston, Leics LE8 1FR
(0533) 884420

An association of British sprinkler companies, concerned to promote greater and more efficient use of sprinkler and other systems using water as a fire extinguishant, and to improve codes and standards by which systems are designed and installed. Represented in the European Commission on the Protection of Heritage Buildings. Willing to advise library and archive staff on the installation, performance and merits of sprinkler systems.

5 BRITISH FIRE PROTECTION SYSTEMS ASSOCIATION
48a Eden Street, Kingston-upon-Thames, Surrey KT1 1EE
(01) 549 5855

An association of manufacturers and installers of fire alarm and fixed fire extinguishing systems. Principal objectives: to improve the quality and performance of fire protection equipment; to promote and maintain contact, and to negotiate, with government and other public bodies, with insurance companies, fire brigades and trade associations; and to publicise the activities of the industry. Willing to give advice on these matters to library and archive staff.

6 BRITISH RECORDS ASSOCIATION
Master's Court, The Charterhouse, Charterhouse Square, London EC1M 6AU
(01) 253 0436

The British Records Association was founded in November 1932 as a national organisation to co-ordinate and encourage the work of the many individuals, authorities, institutions and societies interested in the conservation and use of records.

Its aims are:

(i) Co-operation. The Association serves as a link between all who are concerned in the ownership, custody, preservation, study and publication of records.

(ii) Preservation of records. It seeks to develop informed opinion on the necessity for preserving records of historical importance and the means by which this can be achieved. It will advise on the disposal of papers, deeds and documents of all kinds through its Records Preservation Section. It makes available technical information in collaboration with the Society of Archivists[29].

(iii) Utilisation of records. It furthers the use of records as historical material, particularly through the medium of publication.

Membership is open to any individual or institution interested in records and in the aims of the Association. All who care for the history of Britain as reflected in its wealth of records are invited to join. Subscribing members receive *Archives* (published each Spring and Autumn) without further charge. Correspondence or enquiries concerning the Association should be addressed to the Honorary Secretary.

7 THE BRITISH SECURITY INDUSTRY ASSOCIATION, LTD (BSIA)
Scorpio House, 102 Sydney Street, London SW3 6NL
(01) 352 8219

Founded in 1967, BSIA is an association of companies of repute in the UK which offer security services and security equipment. Its first aim being to promote and encourage a high standard of ethics, service and equipment, BSIA also gives maximum assistance to industry, the public, police and government in the prevention of crime. It represents the security industry on the Home Office Steering Committee on Crime Prevention, and on other government security committees and working parties. Willing to give advice on security matters to library and archive staff.

8 BRITISH STANDARDS INSTITUTION (BSI)
2 Park Street, London W1A 2BS
(01) 629 9000 (Enquiries to (0908) 320066)

An independent organisation and the recognised body in the UK for the preparation and promulgation of national Standards which, in turn, form the basis for a wide range of the industrial and trade Codes of Practice. BSI represents the UK in the International Organisation for Standardisation (ISO), the International Electro-technical Commission (IEC), and in Western European organisations concerned with harmonisation of standards. British Standards are compiled by technical committees, convened by BSI, and consisting of the professions, government departments, user/consumer organisations, manufacturers and retailers. Some British Standards which may not be mandatory requirements may yet be regarded as minimum requirements by enforcing authorities, at local and national levels. Of particular importance in connection with disaster prevention are those Standards and Codes of Practice directly concerned with building design, means of escape, fire tests on building materials, fire-fighting equipment, electrical appliances, and the whole field of building security and alarm systems. An example for instance is BS 4737, which stipulates the British Standard for intruder alarm systems in buildings.

9 THE BUILDING SERVICES RESEARCH AND INFORMATION ASSOCIATION (BSRIA)
Old Bracknell Lane West, Bracknell, Berks RG12 4AH
(0344) 426511

BSRIA is a co-operative research and information association, set up to serve the building services industry. It provides independent, fast, value-for-money services, strictly confidential to the client, all of which are available to non-members on a fee basis. Its information officers are qualified engineers or scientists, and its specialist staff, facilities and instrumentation enable it to offer a comprehensive range of client services devoted to the building services industry. In view of the vast and widespread output of technical literature today, world documentation is scanned and abstracted co-operatively by 14 European countries, for whom BSRIA is the co-ordinating agency. As a result, it provides in this field one of the most comprehensive computerised information services in the world. Of particular interest, perhaps, to library and archive staff, is that BSRIA site teams can conduct tests, whenever possible to a British Standard or other recognised standard, in any part of the country, and on installations of any size or complexity. Such testing can be for proving performance, malfunction investigation, monitoring and evaluating particular factors, and for environmental surveys of temperature, air-movement, illumination, or pollution.

10 DEPARTMENT OF THE ENVIRONMENT (DOE)
2 Marsham Street, London SW1P 3EB
(01) 212 3434

The DOE Building Regulations Division is responsible for making building regulations for England and Wales. Similar responsibility for Northern Ireland lies with the Department of the Environment, Northern Ireland[11]. In England and Wales, local authorities have responsibility for the enforcement of such regulations which are, in part, concerned with safety in fire, and thus of particular importance in disaster prevention. Requirements are given for structural fire precautions which must be observed in the case of all new buildings, in cases of alteration to existing buildings, and where buildings are subject to certain changes of use, including change to use as a public building. Research on safety aspects of fire prevention and protection is conducted by the Fire Research Station[15].

11 DEPARTMENT OF THE ENVIRONMENT, NORTHERN IRELAND (DOENI)
Parliament Buildings, Stormont, Belfast BT4 3SS
(0232) 63210

The DOENI has a range of responsibilities which include highways, traffic flow, traffic signals, the operational efficiency of fire brigades, the provision and maintenance of all public water supplies and sewerage services, and, through the 26 District Councils, the overall responsibility for the administration of local government in Northern Ireland. In addition, it has responsibility for making the Building Regulations for Northern Ireland. Library and archive staff seeking information or advice specifically relating to those regulations should contact the Works Service, Churchill House, Victoria Square, Belfast BT1 4QW, tel. (0232) 234343.

12 FIRE EXTINGUISHING TRADES ASSOCIATION (FETA)
48a Eden Street, Kingston-upon-Thames, Surrey KT1 1EE
(01) 549 8839

FETA provides a central organisation for manufacturers and specialist distributors of all types of fire extinguishers and of other portable fire-fighting equipment and fittings. It operates a joint listing and labelling scheme with the Fire Offices' Committee (FOC)[14]. In order to increase customer/distributor confidence in such labelled products, portable fire extinguishers which are approved by the FOC and distributed by FETA-approved fire engineering companies are provided with a distinguishing FETA/FOC label. It is expected that the range of approval schemes administered by FOC will eventually be taken over by the newly created Loss Prevention Certification Board Ltd[21]. FETA is willing to offer to library and archive staff advice and information on types and distributors of fire-fighting equipment and fittings within their range of activity.

13 FIRE INSURERS' RESEARCH AND TESTING ORGANISATION (FIRTO)
Melrose Avenue, Borehamwood, Herts WD6 2BJ
(01) 207 2345

FIRTO is a test house, with particular expertise in the testing of fire protection equipment, including sprinklers, detectors and extinguishers, and of building materials and structures in respect of their fire properties. Tests are carried out on behalf of government departments, manufacturers, users, and the insurance industry. The existing technical resources and services provided by FIRTO have recently been combined with those of other insurance technical bodies, including the Fire Offices' Committee[14] and the Fire Protection Association (see Appendix C) to form the Loss Prevention Council [22] which will link and exploit them through its new operating company, the Loss Prevention Technical Centre[23], also to be based at Borehamwood.

14 THE FIRE OFFICES' COMMITTEE (FOC)
Aldermary House, Queen Street, London EC4P 4JD
(01) 248 4477

The letters FOC are known world-wide in respect of insurance standards for building construction, rules for the design and installation of automatic sprinklers and other forms of fire protection equipment, and for approval schemes covering such equipment. For many years FOC has been an association of fire insurance companies voluntarily following a code of conduct and practice relating to fire insurance matters, and administering a range of internationally recognised FOC approval schemes. Although such technical activities of the FOC will be continuing, they will now be formally undertaken by the recently formed Loss Prevention Council[22]. Through its newly created subsidiary company, the Loss Prevention Technical Centre[23], the Council will exploit and link FOC resources with those of the Fire Insurers' Research and Testing Organisation[13] and the Fire Protection Association (see Appendix C). The Loss Prevention Technical Centre will also be concerned with the maintenance and updating of technical documents previously issued by the FOC, whose schemes will in future be progessively integrated with those of the new Loss Prevention Certification Board[21].

The currently available FOC rules and lists of approved equipment are as follows:

Rules

1. Automatic sprinkler installations. 29th edition, November 1973, amendments to April 1985.
2. Automatic fire alarm installations for the protection of property. 12th edition. February 1985.

Lists

1. Approved automatic fire alarm systems. July 1985.
2. Approved automatic sprinkler equipment. May 1985.
3. Installers of automatic sprinkler installations. May 1984.

3. Construction and installation of fire break doors and shutters. February 1985.

4. Construction of buildings, Grades 1 and 2. Rev. November 1978, amendments to June 1984.

5. Construction of buildings, Standards I to V. January 1979, amendments to June 1984.

6. Classification of heaters. April 1984.

7. Provision of fire extinguishing appliances. September 1983.

4. Approved portable fire extinguishing appliances. May 1985.

5. Approved drenchers. September 1961.

6. Approved fire break doors and shutters. February 1985.

The sprinkler and fire alarm rules in particular will continue to be actively supported and developed as required by technological changes in their respective fields of fire engineering, and continued support will be given to necessary research initiatives being taken by the Comité Européen des Assurances and others. The recently completed series of experiments on the protection of high-piled storage with sprinklers is likely to lead to considerable improvements in the design of such systems.

15 FIRE RESEARCH STATION (FRS)
Melrose Avenue, Borehamwood, Herts WD6 2BL
(01) 953 6177

Although, as part of the Department of the Environment's Building Research Establishment, the FRS is principally concerned with work related to the fire research requirements of government departments, it also undertakes commissions from commercial or other non-government organisations. Such work is largely conducted within three main divisions:

i. *the Statistics and Operational Research Division* which deals with fire statistics, the economics of fire safety, fire safety strategies, risk evaluation, and the behaviour of people in fire
ii. *the Fire Protection Division* which deals with fire extinction, the identification and reduction of explosion hazards, and the toxicity and other hazards of the products of combustion

iii. *the Fire and Buildings Division* which deals with the fire performance of building materials, components and structures, the spread of fire and smoke, and the overall problems of fire in buildings.

FRS also sponsors complementary and supporting research in universities and elsewhere, and carries out large-scale studies at Cardington in Bedfordshire. Additionally, it produces publications, holds conferences, participates in British Standards Institution[8] work in departmental and industrial committees, and runs an advisory service which could assist library and archive staff concerned with relevant problems.

16 HEALTH AND SAFETY EXECUTIVE (HSE)
Public enquiry points:

i. Baynard's House, Chepstow Place, London W2 4TF
(01) 221 0416 and 221 0870

ii. St Hugh's House, Stanley Precinct, Bootle, Merseyside L20 3QY
(051) 951 4381

iii. Broad Lane, Sheffield S3 7HQ
(0742) 78141

This section of the Department of Employment is the operational arm of the Health and Safety Commission which is responsible to the Secretary of State for Employment, and to other ministers, for the administration of the Health and Safety at Work Act. As an enforcing authority, the HSE is empowered to appoint inspectors for carrying out its enforcement functions. HSE has responsibility for process risks arising from all work activity, including homeworkers (except insofar as enforcement is allocated to local authorities under the Health and Safety (Enforcing Authority) Regulations 1977). The HSE has a duty to provide practical guidance, and (with the Secretary of State's approval) Codes of Practice. The locations of its research and laboratory activities are:

i. Research and Laboratory Division,
Steel City House, West Street, Sheffield, S1 2GQ
(0742) 78141

ii. Explosion and Flame Laboratory,
Harpur Hill, Buxton, Derbyshire, SK17 9JN
(0298) 6211

iii. Occupational Medicine and Hygiene Laboratory,
403 Edgware Road, London NW2 6LN
(01) 450 8911

iv. Safety and Engineering Laboratory,
Broad Lane, Sheffield S3 7HQ
(0742) 78141

17 HOME OFFICE CRIME PREVENTION CENTRE
Police Headquarters, Cannock Road, Stafford ST17 0QC
(0785) 58217

Established in 1963 by the Home Office, at the headquarters of the Staffordshire Constabulary, to provide specialist training in crime prevention for the police services in the UK and for other approved organisations. Facilities include the integrated study of public, industrial and commercial security. The Centre is the national repository for information on vandalism, attacks on safes, and the work of crime prevention panels. A comprehensive index is maintained on security products and services. The aims of the Centre include the dissemination of information, liaising with the security industry, and crystallising thoughts on crime prevention methods. The Centre is willing to give crime prevention advice to library and archive staff, although they are recommended first to consult their local Crime Prevention Office, through the nearest police headquarters.

18 THE INSTITUTE OF PAPER CONSERVATION (IPC)
Leigh Lodge, Leigh, Worcester WR6 5LB

IPC is a specialist organisation, concerned with the conservation of paper and related materials. Its purpose is to increase the professional awareness of the contemporary conservation situation, by co-ordinating the exchange of information and by facilitating contacts between its members, both nationally and internationally. It is administered by an Executive Committee of practising conservators, elected from the membership, together with co-opted technical, scientific, and curatorial advisers. The intention of the Institute is to provide a forum for the exchange of information and a focus for professional awareness that will be helpful to conservators, craftsmen, and scientists, as well as museum curators, private collectors, and commercial organisations. As the principal international body representing paper conservators, the IPC hopes to draw the attention of governments and institutions to the enormous conservation problems of this indispensable part of our cultural heritage. Membership of IPC is open to all those who support the aims of the Institute.

19 THE INTERNATIONAL INSTITUTE FOR CONSERVATION OF HISTORIC AND ARTISTIC WORKS (IIC)
6 Buckingham Street, London WC2N 6BA
(01) 839 5975

Founded in 1950, the IIC is concerned with the whole field of inanimate objects considered worthy of preservation, whether in museums and libraries or exposed externally: their structure, composition, deterioration and conservation. Its members, now in 55 countries, are enabled to keep abreast of technical advances and in personal contact with colleagues at home and overseas, through IIC's congresses and groups, and through its publications, especially the quarterly *Studies in Conservation* and the biennial *Art and Archaeology Abstracts*, which are sent to all members. The Institute is administered by an International Council, drawn from the membership which is open to all who are involved in conservation. With the approval of the Council, Regional Groups with their own structure and by-laws operate in a number of countries. The former American and UK groups have been independently incorporated, while retaining friendly relations with the parent body (cf. United Kingdom Institute for Conservation[30]).

20 THE LIBRARY ASSOCIATION (LA)
7 Ridgmount Street, London WC1E 7AE
(01) 636 7543

Established in 1877, the LA now has nearly 24,000 members throughout the UK and overseas. It is managed by an elected Council, representing many areas of specialist activity and all regions within the UK, and it is the body responsible for approving courses at Britain's library and information science schools, and for conferring the status of 'Chartered Librarian' on those who are appropriately trained and experienced. The LA acts as the voice of professional librarians in matters of public concern, issues guidelines on various librarianship functions, and takes a leading part in the formulation of national and international standards. It provides an extensive range of meetings, conferences, seminars, study schools, exhibitions, journals and books, designed to enable its members to keep informed of developments in librarianship and related subjects. Among its publications are the *Journal of Librarianship* and a monthly magazine, the *Library Association Record*. In conjunction with the British Library, it offers a large library and information service, supplying literature and answering questions on the day-to-day professional problems of its members. The library has one of the world's largest collections of books and periodicals on library science, and provides help with technical queries. In addition to affiliated institutional and other forms of membership, personal membership is open to all who are employed as librarians, bibliographers, or information officers, or who are otherwise connected with the administration of libraries or interested in the objects of the Association.

21 THE LOSS PREVENTION CERTIFICATION BOARD, LTD (LPCB)
Melrose Avenue, Borehamwood, Herts WD6 2BJ
(01) 207 2345

LPCB is a newly developed subsidiary company of the Loss Prevention Council[22]. It will take over the work formerly undertaken by the Fire Offices' Committee[14] and provide an approvals and quality assurance service, to BS 5750, for building construction materials and for the manufacture of protective and warning appliances and devices which prevent loss of life and protect property from fire and other hazards. As part of the government's promotion of standards, quality, and international competitiveness, the National Accreditation Council for Certification Bodies (NACCB) has been formed, and LPCB is seeking accreditation of its approvals and certification schemes.

22 THE LOSS PREVENTION COUNCIL, LTD (LPC)
140 Aldersgate Street, London EC1A 4HY
(01) 606 1050

In view of today's greatly increased potential for disaster, UK insurers have recently taken new initiatives, which are embodied in their setting up the LPC, to link and exploit the existing resources of the Fire Insurers' Research and Testing Organisation[13], the Fire Offices' Committee[14], and the Fire Protection Association (see Appendix C). As a further development, two new operating companies are being formed, to be based at Borehamwood in Hertfordshire: the Loss Prevention Technical Centre[23], and the Loss Prevention Certification Board[21]. The LPC will thus provide industry, commerce, insurers, fire services, architects, and others with a total fire and loss prevention service which will include hazard research, risk assessment, the testing of protection equipment, building products and materials, and information, advisory and consultancy services. LPC is willing to give advice to library and archive staff on appropriate aspects of their planning requirements.

23 THE LOSS PREVENTION TECHNICAL CENTRE, LTD (LPTC)
Melrose Avenue, Borehamwood, Herts WD6 2BJ
(01) 207 2345

LPTC is a new subsidiary company, operating within the recently established Loss Prevention Council[22], and will offer an independent confidential service to its clients in the fields of research, testing and consultancy, based on the combined facilities and resources of The Fire Insurers' Research and Testing Organisation[13], the Fire Offices' Committee (FOC)[14], and the former Insurance Technical Bureau (ITB) which is now no longer an entity but has been absorbed into the Loss Prevention Council. The LPTC will be concerned with long and short term research, with the testing of fire protection equipment and of building materials and structures, and with providing a comprehensive consultancy service on fire and other hazards. It will also be concerned with the maintenance and updating of technical documents previously issued by the FOC.

24 THE MUSEUMS AND GALLERIES COMMISSION
2 Carlton Gardens, London SW1Y 5AA
(01) 930 0995

Originally established in 1930, the Commission advises government and the public generally on museums and galleries, promotes co-operation, and administers a number of grant schemes. It also runs the Government Indemnity Scheme and the provisions whereby works of art may be accepted in lieu of Capital Transfer Tax.

25 NATIONAL LIBRARIES AND RECORD OFFICES
(Conservation Departments)

The British Library	National Preservation Office, Great Russell Street, London WC1B 3DG (01) 636 1544
National Library of Scotland	Preservation Division George IV Bridge, Edinburgh EH1 1EW (031) 226 4531
National Library of Wales	Conservation Co-ordinator, Aberystwyth, Dyfed SY23 3BU (0970) 3816

Public Record Office	Conservation Department, Land Registry Building, Portugal Street, London WC2A 3PH (01) 405 3488
Public Record Office, Northern Ireland	Conservation Department, 66 Balmoral Avenue, Belfast BT9 6NY (0232) 661621
Scottish Record Office	Conservation Department, HM General Register House, Edinburgh EH1 1EW (031) 556 6585

26 THE NATIONAL SUPERVISORY COUNCIL FOR INTRUDER ALARMS (NSCIA)

St Ives House, St Ives Road, Maidenhead, Berks SL6 1QS
(0628) 37512

NSCIA was established in 1971 with the support of the Home Office, the police forces, insurance companies and the security industry. Its aim is to improve the reliability and effectiveness of intruder alarms and reduce the incidence of false alarms. To achieve this, it maintains a register of approved installers who agree to install intruder alarm systems to BS 4737, and who are expected to comply with the November 1981 Code of Practice which supplements and reinforces that standard.

The controlling body of NSCIA is the National Inspection Board which includes members from the Association of British Insurers[3], the British Security Industry Association[7], and various bodies and institutions carefully chosen to ensure that there is no one controlling interest. The inspectorate operates independently, to examine a random selection of installation work for quality and compliance with the British Standard and Code of Practice, and to investigate technical complaints. Library and archive staff in need of more extensive security may obtain from NSCIA a reliable list of the approved installers who will provide a properly designed alarm system, installed to BS 4737, and a round-the-clock repair service.

27 PAPER INDUSTRIES RESEARCH ASSOCIATION (PIRA)
Randalls Road, Leatherhead, Surrey KT22 7RU
(0372) 376161

PIRA is the national technical centre for paper and board, printing and packaging. It carries out research and development projects for the benefit of its members, and also undertakes multi-client research projects for groups of members, and confidential sponsored work for individual firms. Other services include consultancy, technical enquiry, testing, information, training, and techno-economic studies. PIRA is willing to give appropriate advice or information to library and archive staff.

28 ROYAL COMMISSION ON HISTORICAL MANUSCRIPTS
Quality House, Quality Court, Chancery Lane,
London WC2A 1HP
(01) 242 1198

Set up by Royal Warrant in 1869 to enquire into and report on collections of papers of value for the study of history in private hands. Terms of reference were enlarged in 1959, to include all historical records, wherever situated, outside the Public Record Office[25]; with added responsibilities, as a central co-ordinating body, to promote, assist and advise on their proper preservation and storage.

29 THE SOCIETY OF ARCHIVISTS
c/o The Suffolk Record Office, St Andrew House,
County Hall, St Helen's Street, Ipswich IP4 2JS
(0473) 55801 ext. 4239

The Society of Archivists, founded in 1947 as the Society of Local Archivists, has since 1954 been the recognised professional society for archivists, archive conservators and record managers in the British Isles and the Commonwealth. Its aims are to promote the care and preservation of archives and the better administration of archive repositories, to advance the training of its members, and to encourage related research and publication. The Society acts as spokesman for the profession, submitting evidence and comment on matters of professional concern to any official body which seeks advice, or whose activities seem likely to endanger archives. The Society's main publication is the *Journal of the Society of Archivists*, issued twice each year to all members. This carries articles and reviews on professional topics, archive news and reports on Society activities. It also publishes a series of information leaflets on practical subjects, while several of its committees and groups (e.g. the Special Repositories Group, the

Records Management Group, and the Conservation Group) have their own publication programme. All the Society's publications are available to non-members, through the stockholder: David Lamb, Hampshire Record Office, 20 Southgate Street, Winchester, Hants SO23 9EF. Membership of the Society is open to practising archivists, archive conservators, and record managers. The Society will, of course, always be glad to give general advice on archive matters, through its Secretary at the Society's address.

30 UNITED KINGDOM INSTITUTE FOR CONSERVATION (UKIC)
c/o Conservation Department, The Tate Gallery, Millbank, London SW1P 4RG
(01) 821 1313 ext. 295

UKIC represents conservators, and is prepared to point in the right direction people making enquiries on matters relating to conservation (see also the International Institute for Conservation of Historic and Artistic Works[19]).

31 WATER AUTHORITIES

Under the 1973 Water Act, 10 autonomous multi-purpose Water Authorities were created: nine regional authorities in England and the Welsh Water Authority. In its respective area each is responsible for water supply and conservation, for sewerage, for the prevention of river pollution, and for fisheries, land drainage, and the recreational use of water. In Northern Ireland, responsibility for water services rests with the Department of the Environment, Northern Ireland[11], while in Scotland responsibility for water supply, sewerage and sewage disposal is vested in the 12 regional or island councils.

Regional water authorities, England and Wales

Anglian Water Headquarters	Ambury Road, Huntingdon, Cambs PE18 6NZ (0480) 56181
North West Water Authority	Dawson House, Great Sankey, Warrington WA5 3LW (0925) 724321

Northumbrian Water Authority	Northumbria House, Regent Centre, Gosforth, Newcastle upon Tyne NE3 3PX (091) 284 3151
Severn Trent Water Authority	Abelson House, 2297 Coventry Road, Sheldon, Birmingham B26 3PU (021) 743 4222
South West Water	Peninsular House, Rydon Lane, Exeter, Devon EX2 7HR (0392) 219666
Southern Water Authority	Guildbourne House, Worthing, Sussex BN11 1LD (0903) 205252
Thames Water Authority	New River Head, Rosebery Avenue, London EC1R 4TP (01) 837 3300
Welsh Water Authority	Cambrian Way, Brecon, Powys LD3 7HP (0874) 3181
Wessex Water Authority	Wessex House, Passage Street, Bristol, Avon BS2 0JQ (0272) 290611
Yorkshire Water Authority	West Riding House, 67 Albion Street, Leeds LS1 5AA (0532) 448201

Appendix C — The Fire Protection Association (FPA)

THE FIRE PROTECTION ASSOCIATION (FPA)
140 Aldersgate Street, London EC1A 4HX
(01) 606 3757

The role of the FPA is to identify and draw attention to fire dangers, and to provide information and advice on all aspects of fire safety. Its services are specifically designed to assist the fire and security professionals in industry, commerce and the public sector, and their professional advisers: insurance surveyors, fire prevention officers, and consultants. The Association is insurance-backed and is recognised by government, the fire services, and industry, as an independent national fire-safety information and advisory organisation. It operates under a governing Council comprising, in addition to insurers, representatives of government departments, the Confederation of British Industry, and others, and will continue to do so following the setting up of the Loss Prevention Council (LPC) (see Appendix B) and the resultant changes. It will act as publisher for the documentation emanating from the other companies in the LPC, as, for example, the rules and lists of approved equipment and services. In addition to offering information and advice, the FPA provides training for its members at all levels, including the provision of visual aids, and holds seminars, symposia and conferences. It also provides the secretariat service for the Central Fire Liaison Panel, and for the national network of Fire Liaison Panels and local fire protection associations. Although those holding the office of secretary at county level in that network are liable to change, details of the holders in England and Wales and Northern Ireland are given below, as at January 1986, in the belief that they could prove helpful in most planning processes.

FIRE LIAISON PANEL NETWORK: COUNTY LIAISON/ SECRETARIES
(January 1986)

Location	*Secretary*
Avon	C J Leek Imperial Tobacco Limited, P O Box 244, Hartcliffe, Bristol BS99 7UJ (0272) 641111
Bedfordshire	B Coxon Bedfordshire FBHQ, Southfields Road, Kempston, Bedford MK42 7NR (0234) 51081
Royal Berkshire	D J Harper Berkshire FBHQ, 103 Dee Road, Tilehurst, Reading RG3 4BW (0734) 418222
Buckinghamshire	J Hudson Buckinghamshire FBHQ, Cambridge Street, Aylesbury HP20 1BD (0296) 24666
Cambridgeshire	M J Lane Cambridge Fire and Rescue Service, Hinchingbrooke Cottage, Brampton Road, Huntingdon PE18 8NA (0480) 54651
Cheshire	P J Nash Cheshire FBHQ, Walmoor House, Dee Banks, Chester CH3 5UB (0244) 22633

Cleveland/Co Durham	J A Thacker Cleveland County FBHQ, Endeavour House, Stockton Road, Hartlepool TS25 5TB (0429) 872311
Clwyd	DSFPO, F O Jones Clwyd Fire Service HQ, 259 Abergele Road, Colwyn Bay LL29 8AP (0492) 331133
Cornwall	SPFO, K W Riggs Cornwall FBHQ, County Hall, Station Road, Truro TR1 3HA (0872) 73117/9
Cumbria	A P Murray Cumbria FBHQ, Grecian Villa, Cockermouth CA13 0EL (0900) 822503
Derbyshire (no panel) Contact	c/o Chief Fire Officer Derbyshire Fire Service, Old Hall, Burton Road, Littleover, Derby DE3 6EH (0332) 771221
Devon (no panel) Liaison	(SFPO) Devon Fire Brigade P A Kilshaw (0392) 873711
Dorset	D Buswell Dorset FBHQ, North Quay, Weymouth DT4 8TB (0305) 786633
Dyfed (part of West Wales Panel) Liaison	R King (0267) 233481

Essex	A J Stone Essex Fire and Rescue Service, 'C' Division HQ, Fire Station, Rear of Sutton Road, Southend SS2 5TX (0702) 614433
Mid Glamorgan	B E Morgans c/o Perkin Elmer Limited, Industrial Estate, Llantrisant CF7 8YW (0443) 224731
South Glamorgan	K J Wordley Royal Insurance Group, 39 Windsor Place, Cardiff CF1 3SN (0222) 43081
West Glamorgan (part of West Wales Panel)	Secretary: N Walters (0792) 701919
Gloucestershire	Acting Secretary: A R Forber 2 Redcliffe Way, Bristol BS1 6NL (0272) 297478
Gwent (no panel)	
Gwynedd (no panel) Contact	c/o Chief Fire Officer Gwynedd FBHQ, Llanberis Road, Caernarfon LL55 2DF (0286) 3811
Hampshire	M Nash Hampshire Fire Brigade, 'C' Division HQ, Steele Close, Eastleigh, Hants SO5 3AA (0703) 618255
Hereford & Worcester	G G Winkworth FBHQ, Copenhagen Street, Worcester WR1 2HQ (0905) 24454

Hertfordshire	K Woodsmith Hertfordshire FBHQ, Old London Road, Hertford SG13 7LD (0992) 54900
Humberside	D Delaney Norwich Union Insurance Group, Norwich House, Saville Street, Hull HU1 3QQ (0482) 20862 ext. 241
Isle of Man	T E Osborne Tower Insurance Co Ltd, 19 Athol Street, Douglas, Isle of Man (0624) 3446
Kent	R Pearce Kent FBHQ, Tovil, Maidstone ME15 6XB (0622) 54311
Lancashire	J Livesey Lancashire FBHQ, Garstang Road, Fulwood, Preston PR2 3LH (0772) 862545
Leicestershire	D T Bell Leicestershire Fire and Rescue Service, Anstey Frith, Glenfield, Leicester LE3 8HD (0533) 872241 ext. 251
Lincolnshire	R Walker Lincolnshire FBHQ, South Park Avenue, Lincoln LN8 0EL (0522) 33361

London (Greater/Central)	J L McMillan 'H' Division London FBHQ, 90 Old Town, Croydon CR0 1AR (01) 587 4800
London (South)	Divisional Commander LFBHQ Vatric House, 181 Garth Road, Lower Morden, Surrey SM4 4NF (01) 587 2012
London (West)	G Perren 'D' Division FBHQ, Marion House, 61/63 Staines Road, Hounslow TW3 3JQ (01) 572 6246
London (North West)	K Jousiffe BEM London Fire Brigade, 500 Pinner Road, Harrow, Middx (01) 863 1127
London (North)	C J Smith Legal and General Assurance Society, Southgate House, 15 Cannon Hill, London N14 7DA (01) 882 6181
London (North East)	C P Clarke Legal and General Assurance Society, Southgate House, 15 Cannon Hill, London N14 7DA (01) 882 6181
London (South East)	P H Young GI Surveyor, Legal and General Assurance Society, Davis House, 69/77 High Street, Croydon CR9 1EG (01) 686 5595 ext. 236

Manchester	Greater Manchester FBHQ, 146 Bolton Road, Manchester M27 2US (061) 736 5866
Merseyside	J A Warman c/o Commercial Union Assurance Co, Drury House, 19 Water Street, Liverpool L2 0R1 (051) 227 2090 ext. 230
Norfolk	E A Mann Norwich Union Insurance Group, Surrey Street, Norwich NR1 3NS (0603) 622200 ext. 2763
Northamptonshire	L Pritchard Northamptonshire FBHQ, Moulton Way, Moulton Park, Northampton NN3 1XI (0604) 499691
Northumbria	D S Fleming Tyne & Wear FBHQ, Pilgrim Street, Newcastle upon Tyne NE99 1HR (0632) 321224 ext. 275
North Wales and North West Liaison	D H Wesson Royal Insurance Group, P O Box 144, New Hall Place, Liverpool L69 3EN (051) 224 3694
Nottinghamshire	J Toder Gedling Borough Council, Planning & Estates Dept, Council House, Carlton Square, Arnot Hill Park, Arnold, Nottingham NG4 3DN (0602) 269761

Oxfordshire	J W Miller Oxfordshire FBHQ, Sterling Road, Kidlington Oxford OX5 2DU (0865) 4211
Powys (no panel)	D A Edmondson Powys FBHQ, Garth Road, Builth Wells LD2 3AR (0982) 553576
Shropshire	C H Teare Salop FBHQ, St Michael's Street, Shrewsbury SY1 2HJ (0743) 53981
Somerset	N T Smith Somerset FBHQ, Hestercombe House, Cheddon Fitzpaine, Taunton TA2 8LQ (0823) 87222
Staffordshire	C Yearsley Staffordshire FBHQ, Pirehill, Stone, Staffs ST15 0BS (0785) 813234
Suffolk	T J Gedge 'Firepower', Europa House, 14-20 Fore Street, Ipswich IP3 1JZ (0413) 717654
Surrey	R Williamson Surrey FBHQ, 70 Wray Park Road, Reigate, Surrey RH2 0EJ (07372) 42444
Sussex (East)	M Hooper, Assistant Group Officer, East Sussex FBHQ, 24 King Henry's Road, Lewes BN7 1BZ (0273) 473333

Sussex (West)	G P Reid West Sussex FBHQ, Northgate, Chichester PO19 1BD (0243) 786211
Tyne & Wear	(See Northumbria panel)
Warwickshire	A M Mackintosh Warwickshire Fire and Rescue Service, Warwick Street, Leamington Spa CV32 5LH (0926) 23231
West Midlands	R Martin MBE West Midlands FBHQ, Lancaster Circus, Queensway, Birmingham B4 7DE (021) 300 6103
West Wales	SFPO, N Walters West Glamorgan CFS FSHQ, Sway Road, Morriston, Swansea SA6 6JA (0792) 701919
Wiltshire	C Moseley Wiltshire FBHQ, Manor House, Potterne, Devizes SN10 5PP (0380) 3601
Yorkshire (North)	A R Simpson N Yorks FBHQ, Crosby Road, Northallerton DL6 1AB (0609) 779771
Yorkshire (South)	B Copley South Yorks FBHQ, Furnival House, 42/46 Furnival Gate, Sheffield S1 4QP (0742) 27202

Yorkshire (West)	N Byrom West Yorks FBHQ, Oakroyd Hall, Bradford Road, Birkenshaw, Bradford BD11 2DY (0274) 682311
Northern Ireland	W D Jones NFU Mutual & Avon Insurance Group, 4 Adelaide Street, Belfast BT2 8GA (0232) 242961
Belfast Group	S Conn Phoenix Assurance, Northern Bank House, 10 High Street, Belfast BT1 2BQ (0232) 241566
N East Group	A McCrindle Eagle Star Group, 5 Thomas Street, Ballymena, Co Antrim BT43 6AU (0266) 49338
Western Group	R A Allison 21 Dereen Drive, Strathfoyle, Londonderry BT47 1UZ Business: (0504) 61926
Scottish Fire Prevention Council	
Lothian & Borders FPA	F G C Lang Heriot-Watt University, Riccarton, Edinburgh EH14 4AS (031) 449 5111
Grampian	D Dargie Grampian FBHQ, 19 North Anderson Drive, Aberdeen AB9 2TP (0224) 696666

Tayside

G B Ewen
Commercial Union Assurance Co,
Overgate House,
Dundee DD1 1QY
(0382) 25151

Highlands & Islands

G H Grant
Highlands & Islands FBHQ,
16 Harbour Road, Inverness
(0463) 239767

Lanarkshire Group

D Welsh
E Division, Strathclyde FBHQ,
Watermill Hill Street, Motherwell
(0698) 51425

Appendix D — Findings from R & D Report 5925

MAIN CONCLUSIONS
There are a significant number of actual incidents in any year which either threaten or affect libraries, archives and their contents.

Only a small number of institutions have any organised plan aimed at preventing such incidents and effecting recovery from them.

An even smaller number of institutions are in the process of producing plans.

A high level of interest is shown in the value of a flexible master plan.

There is similar support for some form of training to help in the formulation of plans and procedures.

OTHER CONCLUSIONS
There are a significant number of buildings in which periodic fire-drills are not held.

Only about one-third of the buildings have staff trained in the use of fire-fighting equipment.

A considerable number of institutions do not inform the Fire Brigade about their buildings and contents.

In the case of more than a third of the buildings, no contact is made with the Chief Fire Officer.

Almost half the institutions surveyed have no access to equipment for treating damaged books/documents.

Only a very small proportion of institutions have access to freeze-drying, cold-storage, or dehumidifiers; in some of the divisions no such equipment was recorded.

There is no air-conditioning in over half the buildings, and no air-conditioning nor humidity control in about two-thirds of the holding/stack areas.

A significant number of buildings were last re-wired more than 25 years ago, and in a larger number the date of re-wiring was shown as unknown.

There is basement-held material in a number of buildings which are either in flood zones or at risk from water or sewerage disruption.

There are no staff with First Aid experience in about a quarter of the buildings.

RECOMMENDATIONS

Libraries and archives should be encouraged and assisted to promote more widespread disaster awareness and control.

A flexible disaster control plan should be made generally available to those wanting it.

Some form of training and help should be provided for institutions seeking to formulate and understand procedures in:

Prevention (security, fire, flood, storage)
Preparation (disaster planning)
Reaction (response to incidents)
Recovery
Insurance

Other guides

British Library Information Guides are a new series of guides, directories and manuals in the field of librarianship and information science. Four other titles are listed below:

Family directory: information resources on the family edited by Jeffrey Weeks
BLIG 1. ISBN 0 7123 3057 5. 1986.

This directory provides a concise and accessible guide to information resources concerning the family, which will be readily used by researchers, practitioners, specialists and non-specialists alike. Its scope is wide,and though the main contents of this book concern the UK, it also contains a range of references to international resources.

Inventory of abstracting and indexing services produced in the UK by J Stephens
BLIG 2. ISBN 0 7123 3080 1. 1986.

This inventory updates LIR Report 21 of the same title, which is now out of print. The entries are arranged alphabetically by name of service. Four indexes are provided: broad subject headings, specific subject headings, an index of responsible authorities, and an index of database producers with the UK online databases they offer.

Archival collections of non-book materials: a listing and brief description of major national collections by Graham P Cornish
BLIG 3. ISBN 0 7123 3081 X. 1986.

The guide aims to give a description of the major national collections of non-book materials. Materials included in the survey are: sound recordings, visual recordings (film, video, photographs and slides), mixed media materials and graphics.

School librarianship in the United Kingdom by Helen Pain
BLIG 4. ISBN 0 7123 3089 5. 1987.

Using information from a wide variety of sources, including government and HMI reports, local authority documents and previously unpublished surveys, this guide is divided into five main sections covering the current state of school libraries in the UK; the employment of professional librarians in schools; the role of the school librarian; the education and training of school librarians and salaries and conditions of service.